MW00713820

OVERTHINKING

How to declutter your home and your mind to have success in your life

Sofia Tann

© Copyright 2019 by Sofia Tann All rights reserved. This document is geared towards providing exact and reliable information with regards to the topic and issue covered. The publication is sold with the idea that the publisher is not required to render accounting, officially permitted, or otherwise, qualified services. If advice is necessary, legal or professional, a practiced individual in the profession should be ordered. - From a Declaration of Principles which was accepted and approved equally by a Committee of the American Bar Association and a Committee of Publishers and Associations. In no way is it legal to reproduce, duplicate, or transmit any part of this document in either electronic means or in printed format. Recording of this publication is strictly prohibited and any storage of this document is not allowed unless with written permission from the publisher.

All rights reserved. The information provided herein is stated to be truthful and consistent, in that any liability, in terms of inattention or otherwise, by any usage or abuse of any policies, processes, or directions contained within is the solitary and utter responsibility of the recipient reader. Under no circumstances will any legal responsibility or blame be held against the publisher for any reparation, damages, or monetary loss due to the information herein, either directly or indirectly.

Respective authors own all copyrights not held by the publisher. The information herein is offered for informational purposes solely, and is universal as so. The presentation of the information is without contract or any type of guarantee assurance.

The trademarks that are used are without any consent, and the publication of the trademark is without permission or backing by the trademark owner. All trademarks and brands within this book are for clarifying purposes only and are the owned by the owners themselves, not affiliated with this document.

Table of Contents

INTRODUCTION

Before discussing it, it is essential to evaluate every circumstance in your daily life, but we too often swirl about a bath of psychological stress and confusion. The problem of overthinking leads to potential regrets and missed opportunities but can also be prevented from making the most of your life. Here are several steps that can help you to rethink and improve your quality of life.

We must always be conscious of our thinking and actions. Still, a person can hurt himself by overthinking things in a way that leads to self-destructive conduct, to obsessive or compulsive actions. Very often, the individual is not sure how to resolve these problems and can use acts of jealousy and denial, which nourish his insecurity.

CHAPTER ONE

The Curse of Overthinking

Could you overthink what a career-defining presentation could be?

There was a mistake. In one word, indeed.

If you are planning for your next presentation and fear that you may overthink it, try this easy test.

Are you...?

- Obsessing instead of advancing?
- Would you feel like you hit the wall?
- Looking honestly for the magic bullet to get back on track?

Take some time to step back (mentally and physically) and catch your breath if you replied "yes" more than once. You will probably fall victim to one of three common traps. Here are some tips for breaking free and ensuring that you are comfortable, understood, and encouraged as you move forward.

Trap #1: "Spinning the wheels." You repeat the same few lines over and over and struggle with exactly what to say. You probably don't have a consistent core message–a simple phrase that summarizes your entire presentation.

When you ask yourself exactly what your listeners want to say or do at the end of the presentation, you clarify your intentions and quickly get stuck.

Trap #2: "The Curse of Wisdom" If you have read a book Made to Stick, you know what Chip Heath & Dan Heath call the Curse of Wisdom-in an attempt to be complete, we feel compelled to share everything, rather than just knowing what our listeners need to know to be the case. As a consequence, we continue to go... and move... as with the Energizer Bunny and thus stick to our now frustrated audience.

Circle back to this all-important key message-what do I want to say or do when I finish my audience? Identify and not read a laundry list of 3-4 main supporting points. Note that value still outweighs quantity.

Trap #3: "Stuck at the starting door" **Recently,** I helped a client to plan for a webinar that attracts hundreds of people. This lively, concentrated, and inspired presenter quickly stopped trying to create a perfect picture. She was so determined to bolt the ideal icebreaker that she was running out of time and endangered the remainder of her high stakes.

If you're in this situation, open your door mentally and continue. The perfect opening will probably emerge as you create the rest of your post. They may also try to ask a question or cite a startling statistic to capture the attention of listeners from the very beginning.

What is the popular remedy to prevent these three overthrowing traps from becoming a victim? It's never about you–it's about the audience always. Through removing your own mind and switching the emphasis to where it belongs-your audience-you can certainly stop worrying over and begin to move towards a cohesive, successful presentation that delivers your company performance.

Symptoms Of Overthinking

9 Signs You are stuck In An Overthinking Brain. Would you find it difficult at any time to shut down your brain? Were you sick of your thinking and anxious? If so, perhaps you're a chronic overthinker. Unfortunately, overthrowing is becoming a global epidemic, because in tough times we live in which we need so much brainpower. Responsibilities, budgets, emotional trauma, as well as other issues, leave us in an overdriven state. Extensive research found that young and middle-aged adults are particularly involved in overthinking, with 73% of 25-35 years classified as overthinkers. Not necessarily, more women (57%) were classified as overthinkers, then men (43%).

So, if you have difficulty separating and feeling constantly burdened with your thoughts, we have certain tips below to help you stop thinking over, and common signs that it has overcome you.

9 Signs In an overthinking mind, do you find it difficult, at any particular moment, to shut your brain off? Were you stressed and concerned about your thoughts? If so, perhaps you're a chronic reverser. Sadly, overthinking is now a global epidemic as we live in complicated times when we need so much brainpower. Responsibilities, jobs, depression, and other issues leave our minds in an overdriven state. Through extensive research, the University of Michigan psychologist found that young and middle-aged adults were especially overthinking, 73 percent of 25-35-year-olds being identified as overthinkers. It is not surprising that more women (57%) than men (43%) identify themselves as overthinkers.

So, if you have problems with detaching your thoughts and being continually burdened by them, we have some tips to help you stop thinking and popular signs that your thoughts have overtaken you.

9 SIGN YOU SHORT IN YOUR MIND

1. INSOMNIA REGULARITY Overthinkers are familiar with the difficulty of falling asleep. Insomnia grips you because you can't seem to turn your brain off, and your thoughts paralyze you slowly. Your mind races and you feel too close to sleep; every concern of the day keeps your mind inundated, and you can't escape the mental jail.

9 Signs You are stuck in an overthinking mind. Do you find it difficult to close your brain at any time? Were you stressed and concerned about your thoughts? If so, perhaps you're a pathological overthinker. Unfortunately, rethinking has been converted into a global epidemic, because we live in dynamic times, in which we need so much brainpower. Responsibilities, finances, emotional trauma, and other problems leave us in an overdriving state. Through extensive research, the University of Michigan psychologist Susan Nolen-Hoeksema found that young and young adults are particularly involved in overthinking, with 73 percent of 25-35 year-olds identified as surgeons. Not unexpectedly, overthinkers are more women (57 percent) than men (43 percent).

There are, therefore, some suggestions below that will help you to avoid overthinking, and through signs that your thoughts have consumed you if you have problems getting rid of your thoughts and feel constantly frustrated by them.

9 IN YOUR MIND YOU SIGN IN YOUR MIND. REGULAR INSOMNIA Overthinkers are all too familiar with the

problem of falling asleep. Sleeplessness captures you because you cannot seem to turn your brain off, and your thoughts slowly paralyze you. Your mind is racing, and you feel wired too much to sleep; all the concerns from day to day flood your mind, and you cannot escape this mental prison.

Try to do relaxing things in front of the bed like yoga, meditating, drawing, coloring, reading, writing, or even talking to a loved one. Do something that takes your mind off and something else to expose your imagination and emotions.

2. If you live in fear of the future, you are certainly trapped in your mind. LIVING IN FEAR Nolen-Hoeksema found that, in her study, this apprehension causes reverse thinkers to turn to drugs or alcohol to suppress their negative thoughts.

You should start a meditation practice or another activity that encourages mindfulness and living in the present moment for this overthinking symptom. One idea is to create a "window" to overthink. Enable 15 to 30 minutes a day, whether by writing or speaking to someone, to get rid of all your worries. You should start with your day and place the worries in the dirt.

3. Overthinkers have a major problem: they have a need to monitor everything. Overthinkers have one main problem. We want to prepare the future, but this gives them great anxiety because we can't predict it. They don't like anything they can't control. You have a great fear of the unknown, which causes you to sit and think about all options rather than take action.

Yes, this UC Santa Barbara study found that overthinking leads to bad decisions and decisions.

Try to bring yourself back to the present moment with deep breaths and focus on something that relaxes you. Try to think about how these thoughts will benefit you right now, and it should be done by yourself since you will find that they do nothing for you but cause great pressure.

4. FEAR OF FAILURE Overthinkers also has a continuous desire for perfection. They can not accept failure and do all they can to avoid it. Ironically, it usually does not involve doing anything. Remember, the overthinker paralyzes fear, and instead of risking failure, they will never be able to fail.

If that sounds like you, note that you are far more than your mistakes and failures. However, keep in mind that you have to make certain mistakes to get anywhere in life. You can evolve, learn, and reach new heights in your evolution.

5. CONSTANTLY YOURSELF SECOND-GUESSING. Due to their eagerness for perfection, overthinkers continuously reanalyze, analyze, and double analyze every other situation. You don't want to make the wrong choice, so you take a very long time to make that choice because you don't trust yourself. We are untouched by their instincts, so every choice comes from the brain, and it's not always healthy. When the brain is so nebulous and tangled that you can't make a clear decision, you definitely are an overthinker.

Learn to trust your intuition and go to your intestines. If it turns out to be negative, you will, at least, have learned from the experience and have more life lessons.

6. FREQUENT HEADACHES You probably think too much, if you have daily headaches. Headaches signal to our bodies that we need a break, and this includes rest from our own minds. Even if you pay close attention to your ideas, you will still think the same things.

Worriers appear to have pessimistic patterns of thought in a loop, but then aim to reinforce optimistic thinking to counteract this. Please spend time on your breathing and attention, and notice that headaches are gone in no time.

7. STIFF JOINTS AND MUSCLES. Believe it or not, it may affect the entire body. When something affects your physical body, it travels into your mental body, and until the root problem is addressed, you will continue to suffer from grief and pain. Overthinking can start in the brain, but its effects infiltrate other areas of your body, making you feel tired and lethargic.

Try to stretch before bed every night and get regular exercise. This should help to develop a healthy lifestyle, and perhaps, a healthy mind. The mind and body interact very closely; if one is out of balance, the other would always slip to the side.

8. FATIGUE If we feel constantly tired, we need an action plan. The bodies want us to tune and listen to their cues rather than continually go from one task to another and ignore their calls. While too much exhaustion can be induced and not rest, rethinking can also lead to fatigue. Think about it (but of course not too hard): you don't give your mind a rest when you always think about things. Your mind can't run 24 hours a day, and you're going to get burnt out.

We didn't have much to worry about when we lived out in nature, so we had less to think about. In the modern world, we have complicated lives that allow us to do so much in such little time, but that is why we need to slow down even more and take greater care of our well-being. If you feel tired, slow down and understand what your mind and body need.

9. Failure TO STAY Throughout the PRESENT Point in time. You can not even live in the present time and enjoy life, as it happens. Too much thought causes you to lose focus and get

lost in your head. Getting stuck with thoughts takes you away from now and may disrupt your relationship with others.

Try to open your heart and mind to the world around you and not be so stuck in negative thoughts. Just encourage your thoughts to serve your well-being and try to ignore the ones that only bring down you. But you can only understand this if you know how to balance your brain and heart instead. Life offers so much beauty and the potential for incredible experiences.

When we pay attention to others, we take a break and therefore concentrate on another human. Learn to listen to others, communicate with them, and ask questions about your life. Through developing groups and learning to collaborate and communicate, we can avoid this persistent overthinking issue together.

In short, concentrate on doing people who make you feel good and inspire you to keep active. Start your training program, join groups in your community to connect with like-minded people, eat healthy foods, practice consciousness, and, most importantly, learn to maintain positive relationships with yourself. See your thoughts as instruments for your development, not as enemies that hamper your progress.

Explanation Of Overthinking Disorder?

Overthink Depression-What's that?

There is no overthinking illness. Many different types of anxiety disorders are associated with an individual's rethinking or rumor, but no disorder exists. If a person cannot stop being obsessed and concerned about things, they can impair your quality of life.

Several conditions for mental health where a person can't prevent rumination include PTSD, anxiety disorders, depression, panic disorder, agoraphobia, separation anxiety disorder, selective mutism, phobias, social anxiety disorders, or some other condition may be a symptom.

Many of them have overthought as a symptom when it comes to anxiety disorders. For instance, a person with a panic disorder could ruminate and overthink again when a panic attack occurs. They are obsessed with something that might cause their assault. They are not only anxious, but they also have meta-anxiety, which is anxiety. A panic attack overthought made it more overwhelming.

It's normal to overthink. You don't need an anxiety disorder to ruminate continually. You might argue that it's part of the human condition. Sometimes we all overthink things: you can be too concerned about what you said or did to someone. You can be concerned about school or work results. You may be worried about how others see you. These are all examples of how you could rethink.

Certain instances of overthinking include:-If you are concerned about what you should have said, or done-performance issues or worried about comparing yourself to others at work–getting into "what if" situations in which you wonder what might be happening in a number of circumstances; Many people suffer from obsessions and worries about things outside their control. Cognitive-behavioral therapy (CBT) is a common treatment for this form of anxiety. CBT helps people question their negative or irrational thoughts and turn their thought into productive and positive thinking. Therapy or panic therapy will make a huge difference for someone who worries twice. You can work here at BetterHelp with a therapist in your area or with a trained mental health professional. Digital therapy is an ideal place to work on depression and to learn how to cope.

Overthinking Most people understand the term anxiety disorder (and, yes, millions of Americans suffer from a sort of anxiety disorder each day), but we tend to ignore a significant symptom of overthinking anxiety disorder.

The definition of rethinking is something to ruminate or obsess about. Some people might believe that they are overthinkers when they hear this description. Who doesn't go without thinking about something for one day? We wonder whether we make the right choices from small things such as choosing the fastest route for our tour that morning or choosing the right restaurant to dine in things like the well-being of our children and the safety and security of our family. But that's usual. That's normal. It is common to worry and to some extent, to overthink.

Nevertheless, overthinking can have harmful effects on a person mentally and emotionally. If we overthink as to an anxiety disorder, we will think excessively about something which causes anxiety, stress, fear, or fear. It isn't just worrying too much about something–it's so obsessed about something that this affects your ability to work in your life.

If you ask or worry about yourself, your life, your family, your friends, or anything else, and you don't have a thought-out problem, you worry for some time, and then you go on your day after a short time. Sometimes you continue to worry, but you don't ruminate constantly. You find that the concern doesn't interfere with the rest of your life. Furthermore, with the overthink of an anxiety disorder, the problem is that everyone can think about it, and even if they don't obsess with the same thing all of the time, they are always nervous.

When you think that you might get overthrown because of fear, you might have found that you have encountered one or more of these situations:-Following up with and assisting in conversations as you keep going through possible responses or comments until the talk is over or the opportunity to speak is lost- Nevertheless, those who experience this find that their quality of life is impaired by their inability to control negative thoughts and emotions effectively. It can make it harder to socialize, enjoy sports, or be productive at work because your mind invests excessive time and energy in certain lines of thought. There's a sense that you don't have full control over your own minds or emotions, which can harm your mental health.

It may be difficult to get friends or to keep them, whether you fail to communicate when something is wrong or because you can overly communicate. It can be very difficult to talk to them, and you think about what to do or do with them because you are too concerned about how you are going to do or what is going to happen. Someone who overthinks can even struggle in a general conversation or in a normal environment. You may have trouble even going to the store or to an appointment.

How to stop overthinking

The reality is that overthinking will influence everything in your life. It can affect the way you communicate with others, impact your social life, and make a difference to your personal life. What it all means is it could begin to wear away yourself and your relationships with the someone around you. Overthinking could make your life seriously.

How do you stop thinking, "Stop thinking about things!" You may have heard this many times, and it's very helpless. You can't just change a button and stop thinking. Yes, being asked to stop thinking sometimes ends up thinking more about you. This is a vicious cycle. It is a cycle.

Essentially, it's a long process that you need to train your brain to try not to overthink. Let's look at some common reasons for overthinking and showing you how to avoid overthinking.

Insomnia Overthinking

Your mind will race, and you can have obsessive thoughts about getting to sleep when you can't sleep. Such overthinking also occurs when insomnia hits, and the next day begins. You can feel tired and less concentrated on your brain. You may have negative thinking and obsessive thinking that you can't sleep.

Distractions in anxiety and overthinking are always important. Why you should start paying attention to your issues, distractions will reduce your anxiety, stress, and other problems. Try to watch a movie or work on a puzzle.

Start to notice when you have an anxiety attack rumblings. Then try and get out. In many cases, anxiety and rethinking can be avoided, especially if you know the triggers.

If you're too nervous and over-thinking and taking all your time, it's time to see a doctor.

Anxiety and rethinking are two things that go hand in hand, but you can do much better by managing your fear and rethinking.

When you think about bipolar disorder, you tend to think about the information you know about mental health. And that health information is because people with bipolar disorders are either depressed or manic. People with it have difficulty with their mood, but they also have difficulty thinking.

overthinking and bipolar disorder

With overthinking and bipolar disorder, disturbing or troubling thoughts can occur on both sides of the coin. One may be worried about what will happen in the future with depression. Alternatively, they may think about the medication's side effects.

In mania, you can find it difficult to pay attention to your ideas, making it more difficult to question your thoughts. It is sometimes difficult to separate real life from fiction. And, maybe you are so euphoric that you spend time to feel safe and then regret it.

It is important that you seek the help of online therapy or a personal therapist with bipolar disorder. Online therapy is particularly effective moderate cases in mild. A therapist can provide you with basic information about bipolar disorder as well as information on mental health. Therefore, it is important that you take care of your thoughts.

Your bipolar episodes can sometimes last for various periods, and your thinking may worsen them. You will dwell on the negative and exacerbate the condition for long periods. The emotions that come into the heads continue to exacerbate the problem, bipolar disorder, or not. Seek assistance if you need it.

thinking positive thoughts.

You might wonder how to stop thinking positive thoughts. One way is to think about more positive ideas. You can roll your eyes. You think positive thoughts are something straight out of a cheesy health book, by definition.

Nevertheless, science supports positive thinking and more optimistic forms of thinking. If you want to think better, here are a few ways to do this.

See your confirmatory bias. Negative thoughts tend to linger, and it generally is the reverse when it comes to positive thoughts. You should try to change the way you think.

Start noticing the positive thoughts instead. Write it down when you start to think positively. Note when you are dreaming and collecting positive thoughts.

We can't emphasize this enough. Practice attentiveness. Cautious techniques teach you to relieve any emotional distress. Destructive thinking passes through the window. Concentrate your attention on positive ideas.

Think of all the times that you have helped people. Think of something that is perfect for you. Just let any distressing feelings go. It is a step process requiring practice.

Some people believe that positive thinking does not mean negative thinking. Both matters should be overlooked, no matter how great they are. This is not at all real. To think positive thoughts means to think less negatively in the rethinking department. There will be distressing thoughts, but positive thinking shows that emotional distress is temporary and that there is much to think about.

Try to clean up your social media feed. Delete the more negative people and focus on positivity. Yes, you ought to eat negative news too, but many are overloaded, and it isn't good for you if you are an overthinker.

Mentally Strong People

It is less likely to overthink mentally strong people. View the brain as a muscle. The more you train it, the better mentally you get. It is especially important as you age to increase your mental health strength. Mental health decreases with aging, but you can train the mind with the right health information.

Here is some good information for your mind about mental health:

- Mentally strong people practice a lot. You can imagine strong people strengthening your body while thinking of exercise. Your mind, however, is affected by many positive side effects. Your brain, for example, releases good feeling chemicals that kill pain and help reduce stress hormones. Furthermore, exercise distracts you from your thoughts and makes it great if you want to know how to stop thinking over.
- Mentally strong people try as much as possible to socialize. Try to talk to a friend and reach them deeper. If you do not have friends to talk to, try going to a bookstore, a café or other place to talk to someone. You are much less worried and more relaxed as you talk to new people and try to make friends.
- Daily behavioral therapy is done for mentally strong individuals. A type of therapy helps you get rid of misfits and thoughts and can be used to treat mental illnesses of all kinds. Eating disorders, bipolar disorder, pervasive depression, and more.
- Strong people prefer to prepare by mixing it up physically. The same stuff can have negative side effects over and over. Look at your life and think differently about what you can do. Try a new hobby, go to your dream job, or learn something new. When you begin to live for a new day, it helps you think over.
- Strong people need to understand the moments of vulnerability will occur. Often you spend too much time thinking, and then you see yourself overthinking.

It will happen, so you can't overthink this. It happens sometimes. Don't spend hours there. You should schedule a time to wander around a certain problem and stop thinking about it when that time is over. This can take place in training. However, powerful people can certainly seek it.

Stress Reduction

You may wonder why "stress reduction" is here. Well, our tendency to reverse stress goes hand in hand. Stress is our body's way of helping us when we are in a situation that threatens us. However, our body can't distinguish between real danger and common problems and thus the amount of stress.

It is difficult for people to cope with all their stress.

Stress may be good. Stress associated with positive, stressful psychology tends to challenge and to improve you. Nonetheless, only positive psychology goes so far. Excessive pressure will exacerbate your issues, including:

- Making you fear rejection, culpability, failure, or defeat.
- The things you can't control tend to worry about it. Most people know that they should be worried about things that they can change and that they can't change things, but it's difficult to overthink that.
- The physical stress is too stressful a problem. Hurts every day, literally. Physical stress indicates stress seems to have painful and effective effects. Instances of physical stress provide headaches and other body pain, possibly clinical depression as well.

Anyone can be anxious. Whether you are a boy, a teenager, or an adult doesn't matter. If you are used to overthinking and worrying, here are several easy ways to reduce stress. Anyone could do these creative ways, and neither does a doctor need such simple ways.

- Practice behavioral, cognitive therapy. There's something that takes some practice, but it is important to learn to recognize thoughts that are distracting and, by extension, to cope with them.
- Write down and arrange the issues from the most critical to the least. Part of troubleshooting requires first and foremost solving the easiest problem. Eventually, you will find easy troubleshooting.
- Think of your fear of insufficiency and other everyday fears. Why are you afraid of this? How does your stress affect you? Are you afraid of guilt or failure?
- The importance of working out isn't understood by men. It can help reduce anxiety when looking for something.
- Take a while to relax. See your favorite show for what's happening. Don't spend too much time stubbornly, but rather take a break, return with a fresh mind.
- Do not smoke or drink alcohol. If it encourages them to prescribe medication with a psychiatrist or counselor, take it.
- Seek to work with a therapist finally. We will help us with your problems.

Hypochondria

Hypochondria is also a part of rethinking and general anxiety disorder. That's when you still worry that you have something wrong clinically, making it a big problem to reconsider.

Many people are slightly hypochondriac. For example, when you visit Dr. Google, you may think something is wrong with you, normally then you talk to your doctor. Instead, you find that nothing is wrong and that it is just a question to worry about, combined with some generalized anxiety disorder.

You might be a severe hypochondriac, though. You also see your doctor today about something. You still have these feelings after speaking to your doctor, and no discussion with your doctor seems to make them go away. You still think you're sick, no matter how much you try.

That's something you have to look for. You may not have only a generalized anxiety disorder. You will finally work up the courage to admit that you are good after having counseling.

Seek Motivation

Even though many people are skeptical of motivational speakers, they can help. Hearing stories about such a man who could overcome anxiety and live or learned to live in an older age can inspire you and is a good way to distract you from thinking again.

It's a good way to get some personal information about mental health. Although certain of this information on mental health may not be part of contemporary psychology, some are worth checking out.

A spiritual psychologist, for instance, is a good place to get information about health. "Eckhart Tolle, religious psychologist, and writer," you ask.

Guy Winch, a psychologist, is also a good listener. Every author of emotional guidance should be examined.

Read all other medical information you can get when it comes to awareness. Some books are short, and it doesn't take long to read. Someone else takes a long time, but it's worth the advice they give.

It is important to consume all your awareness material when it comes to overthinking. Consciousness is the key to getting the assistance you need.

Anything Else?

It is worth noting that it is a mystery how we feel as well as how the brain works. Many clinical trials, both social and clinical, can teach us something about the mind. These clinical trials, however, are only social, clinical, and mental.

Perhaps one day, we will have a pill to fix an overthinking, which is not reserved for clinical trials but is a long day away.

Overthinking is an attitude that can always occur. It can also have the symptom of overthinking for someone who has anxiety or any kind of anxiety disorder. The fear and concern about different circumstances and barriers of your life will easily become worried about what you need to do or how bad things might be avoided. The truth is, you cannot stop all the bad things, and you can't stop any bad decisions from happening. What you can do is get assistance.

It may be helpful to seek professional treatment if you have struggled to stop overthinking. Help can be found in many ways, but a convenient, private place to start is via an online advice site such as BetterHelp.com. Here you will find access to approved advisors to assist you in managing your challenges. You don't have to stop your own thoughts. Trust an online therapist to help you think better about your life and live your life in healthy ways every day.

You can connect with a certified, private mental health provider through online therapy without thinking about going to an institution or even being seen by anyone other than the therapist. You should feel more comfortable because you feel better in your house. Not only that, but you will have power over what is happening. All these things can make it easier for you and your therapist to begin your healing journey. You just have to find the right one.

If you are nervous, there is hope. Your widespread anxiety disorder does not match a therapist. We struggle with common anxiety disorders, depression, bipolar disorder, and other overthinking issues. You should work with a professional mental health consultant to help you develop the ability to cope with your depression and lead a fulfilling life. Check below for some therapists ' comments.

Insomnia is for a reason called a vicious cycle. If you have it, it is difficult to stop worrying about not sleeping. Here are a few ways that you can reduce this to sleeping difficulties.

Games for relaxation and concentration. These allow you to live in this moment, to remove intrusive thoughts or emotions. In addition to the fact that focus can train the brain, it can calm you down and make sleeping easier.

When you can't fall asleep, get out of bed. If you are in bed and you can't go to sleep, it feels like you can not go to bed. The subconscious connects restlessness with your heart. Go out and chill. Go out. Don't waste time on social media or do distracting activities. Calm instead. Relax.

Realize you won't die because of a lack of sleep. While your fear and fear tell you not to sleep, most cases are temporary. When your health decreases because of chronic sleep deprivation, thinking too much about sometimes insomnia can make the problem worse. If the problem persists, a doctor or a psychologist must be seen.

Making a decision Another reason people are overthinking is because of the decision-making process. Sometimes it's a big decision. Some times, decision-making is a bit stupid, like what to choose from a restaurant.

While you should think about your decisions, just overcome them, particularly when you are waiting for a lot of people to decide. Here is how to eventually conquer uncertainty around decision-making.

Decision-making

Allow the decision-making time limits. Now, this time limit must not be so short that you feel extremely long, but it must be short enough to help you stop thinking.

A large number of people, especially the white decision-makers, plan their thinking times and in the meantime, get distracted. Sometimes thinking can prevent overcoming anxiety.

Again, attention and the four noble truths could help. The human condition should include the logic behind the decision, not the unfounded fears.

You could change your opinion on a decision in some situations. If this is known, it can make a decision easier.

Overthinking and Anxiety

Some mental illnesses could lead to overthinking and an apparent association between anxiety and overthinking. Those who are nervous never live in the present.

Some of the brains are always worried about what is next, and extreme anxiety and disorder can make it difficult for you to leave your home.

Here is how to stop the fear and overthink when your anxious brain says no. If you set goals too large, your anxiety may overthink things. Fear and anxiety make it difficult for you to set larger goals. You will work your way up by setting smaller goals.

Once, we cannot emphasize the importance of reflection and concentration. It is good for many mental illnesses, particularly when anxiety and overthinking are at stake. Meditation can pull you back to this moment and calm your body if anxiety strikes.

Find out what will overthink your nervous brain. Causes can make your mental illness worse and make it easier to handle by writing down what causes your depression and overthink.

Some Effective Steps of Overthinking

It's very comfortable to fall enter a trap of surmounting minor something in life. So, when you stop and think about something, ask yourself simple questions. It was found through research that broadening the perspective can quickly get you out of thinking with these simple questions.

Try to set short decision time limits. And learn to make better choices and take action by setting deadlines in your everyday life. No matter if it's a small decision or a bigger one.

Be an action person. When you learn how to start acting correctly, then you will do less by overthinking. Setting deadlines is one thing that will help you to act.

Try to realize that you can't control everything. Trying to think about something 50 times may be a way to try to control everything so that you can't make a mistake, fail or look like a fool. But these things are part of a life where you really extend your comfort area.

Say stop when you know that you can't think straight. Sometimes if you're hungry or lying in bed and about to sleep, negative thoughts start to bustle in your mind.

Do not lose yourself in vague fears. Another trap that you've fallen into many times is that you have lost your fears about a situation in your lives. And so your wild mind created disaster scenarios of what might happen if you do something. What's the worst possible? You should learn to ask yourself this question.

Do spend most of your time now. Be in your everyday lives at this moment rather than in the past or a possible future. Right now, slow down how you do anything you do. For example, move slower, talk slower, or ride your bicycle more slowly. It makes you more aware of how you use your body and what is happening around you.

CHAPTER TWO

How to Form Successful Habits

You have chosen to create a new habit to cut back your recent over-eating indulgence. The children have even commented and noted the' winter weight,' which has recently slowly crept on you. The café you sneak in during the day is an unnecessary luxury. You've granted a defeat-it's time to trade the gym membership coffee card.

It has been said that the creation of new behaviors is difficult because it disrupts the natural state of equilibrium of both mind and body. Although the rational mind has easily stated a clear YES for the new habits, the emotional brain is not as excited about your new plans.

You have thought about it with every luck, together with your strong emotional desire to change it. We seem to have little idea of the fight ahead until we begin work on it. In my early adult life, I was at the mercy of my patterns because I was vulnerable to my emotions. Due to unreasonable expectations on my part, a variety of well-meant practices were met with resistance midway.

As I am approaching medieval age, I have had the chance to develop healthy habits in various areas of life that continue to serve me well. I do assume my clients have benefitted from my wise therapy and steep learning curve over the years as a healthcare and self-empowerment practitioner.

I would like to outline five key points of value for the formation and maintenance of new practices. In conjunction with your daily routine, they shape the underlying desire for lasting change.

1. Know the cycle of change: I've been working with a sports psychologist recently to understand the importance of the cycle of change in developing new habits. Undoubtedly you will experience inner resistance when you adopt new habits because you disrupt the stability of mind and body. Understanding six stages of progress in advance help you to foresee the journey ahead confidently. A related piece of excitement: 33 percent of people canceling or occasionally attending a fitness membership after the third month. Knowing that motivating habits of people are waning over time, fitness centers deliberately encourage you to sign 12-month contracts paid in advance with tiny exit clauses.

2. Have an impressive reason: avoid beginning a new habit with the belief that it is right. Remember that despite the best intentions, the conscious and emotional brain has different agendas. You will certainly have resistance as things get harder because internal conflicts are bound to arise. It is best to use a conscious goal to follow the new habit. Motivational Jim Rohn once whispered, " one of two things we must all endure is the agony of restraint or sorrow or disappointment." We all know that discipline is lighter than regret.

Back to your WHY? Help you interact with your mental and aware thoughts. It is crucial to succeeding if you connect with your original intention to start a new habit. As the path gains momentum, delays and internal opposition are often necessary to slow down the progress.

3. Chunk it down: split the objective down into smaller objectives. Follow one rule or target at a time until you are an expert in it. For example, if your intention to' go fit' means a practice program, you may start by testing the waters early in the morning with a series of long gentle walks. Don't expect creative forms your wellness plan will take shape. Beginning slowly with the intention of gaining momentum, in the long run, maybe much more useful than stopping. Allow the force of the target or habit to drive you into action. As the saying goes, the race wins slowly and steadily.

4. Manage your environment: delete temptations that could ruin your development. If you are new to reducing unhealthy foods, make sure that your refrigerator and cupboard are filled with healthy food options. While it may seem insignificant, the conscious brain is irrational in times of emotional need, which contributes to the possibility of cheating. Hold temptations out of sight everywhere you can. Likewise, avoid falling into the lure of being rewarded with food. The mind is incredibly astute to understand it, having undergone thousands of years of evolution-you must find ways to use the bonuses. Choose for non-food benefits such as massages, the purchasing of a new item, music, and so on. It is important to resist the formula as you will always get unstuck at times. Don't be harsh on yourself when this happens. Use your time wisely to refine and keep following your tradition.

5. Commit to the custom: time to put the metal pedal on! Smaller wins early in the habits give the habit a critical boost. Daily activities for a whole month are a timely way to build sound disciplinary behavior. Regular action is essential to

maintain momentum rather than sporadic application. Target at least 90% + strike rate in the first month. I consider it useful to use a variety of tools as incentives. I purposefully put colored post-it notes around the house where I often go. If you use technology to motivate you not to fall victim to the technology, use it instead to help you keep to your new plans.

In the end, setbacks are sometimes inevitable throughout the habit formation period. Make a public statement to a friend, colleague, or loved one of your desired customs. Take responsibility for someone who is likely to offer much-needed support or who has walked in your shoes. Offer to give the favor back. Responsibility for someone gives you a good reason to keep your word. It makes it all the easier to stick to your habit.

Before going, fail to overthink your feelings or fall victim to them as this becomes difficult. Naturally, your mind will find excuses to jeopardize your progress. Don't buy in lies.

Get Rid of Overthinking Once and For All!

Listening to the voices in your head is so easy. To be seduced by their analytical allowances. They don't leave any unturned leaf. They guide you through the turbulent waters of fear and insecurity and ultimately let you do... Nothing at all!

People sometimes overthink circumstances and create scenarios that don't exist in their minds. We try to convince themselves of a certain situation so that for fear of criticism, we will not risk taking responsibility for their actions. Or make hypothetical conversations and events to support any insidious thoughts and feelings in relationships, at work, or with family.

We must always be mindful of our thoughts and actions, but a person can have an adverse effect on oneself by rethinking things in a way that leads to self-destructive conduct or delays. The individual is often not really sure how to solve these problems and can use acts of envy and degradation that foster their insecurity.

The retraining of one's mind is very complicated so that one is not eaten with meaningless thoughts that have no real purpose. You must first recognize who you are. I know that sounds simple, but if they have no real sense of identity, they will feel less than many things in life and are threatened as an individual. Recall, how people are seen represents how they see themselves. If someone has no self-esteem, others will see. A job loss or a relationship breakdown can cause many of these sensations.

Additionally, focus on the good things in life and remember them. Sometimes you may like there's no. But there are. There are. Avoid being so critical, therefore. One must remember all the good they did, the promises they kept, and the friendships they kept. It's healthy to balance the images of a person and the real person below. Feeling good, in turn, nourishes the soul, and positive and productive thoughts and actions are manifest.

Thirdly, plot a course, and do it. Hands of idleness stir dumb minds. A person might plan the next step in a lifetime and time again, but only then they will overthink. Taking small steps towards the target makes them possible. When you do this, you feel good about life, creating harmony and self-worth within your mind and having value as a person.

To overthink can be a toxic and weakening mental exercise that prevents a person from making sound decisions and impairs his judgment. A stronger sense of self-worth will not only encourage you to speak in more optimistic terms, but it will allow you to break out of the prison in which you have put yourself and refuse to love, happiness, and success.

Overthinking Will Cause You to Think It Over

When was the last time you did anything without spending a second considering whether it is right? When, in the simplest things, you find it quite difficult to decide what to eat for breakfast, you may be a victim of what we call "overthinking."

A typical rational person has a gift of rationality that helps him to know the difference between good and bad. "reason" often causes us to worry too much, causing us to be nervous in all our decisions. Having a hidden meaning in circumstances that do not require your work concerns your mind that in everything you do, you continue to become conscious, limiting you to do what you want. Each choice weighs overthinkers before they can pass. Even though they know that they are good, they tend to ignore the larger picture that causes them to panic under pressure.

As we grow up, we are filled with so much learning that, to some degree, we gain so much good knowledge.

If you're not in the mood, you appear to be anxious and depressed. This will potentially stimulate your mind into thought. Sometimes these feelings don't even apply to the main reason you're in a bad mood. If you begin to think too much, it affects your normal functions.

Stop looking at what is beyond the horizon if you don't have to. You must have a clear view of what lies ahead before you dig deeper. You will fail to see the truth of all things if you keep giving weight to those who are unnecessary.

In a continuous cycle, once you start thinking about something, ask yourself: is it still important in the long term?

If you overthink, you should ask yourself this simple question immediately. Then you can spend your attention on something that really matters. Prepare to set a time limit for all your decisions so that you can act on them immediately. Make decisions easier by setting "deadlines" in life to discourage you from moving on and motivate you not to take things for granted.

Most of us would like us to control all that happens. If this were only feasible, we could have prevented failure, and things would have been fine. Yet mistakes are the only way we can learn a lesson and become a better person. Everyone who succeeds made mistakes before reaching their status quo.

Stop thinking all the time... It's only going to be an endless cycle, and none of it matters! You can't control it all. Accept the fact that there are only issues beyond what you can deal with.

Four Tips to Confront Negative Mindsets

A few authors also experience negative thinking throughout their writing careers. They have become accustomed to criticizing themselves through negative self-speech. Such negative thoughts can lead writers to unproductivity and even hopelessness.

Particularly for writers, it is important to avoid these negative feelings because, when they are scared, they can not be productive and happy authors. Many times we get the best of these negative statements and can not produce the best manuscripts. We can be most effective if we can suppress these thoughts.

This is why it is so important for writers to turn from negative to positive. In the following, I outline four steps to change negative thinking most easily.

1. Do not overthink.

There is nothing more depressing or pessimistic than to overthink. Do not overthink. If you overthink, your brain is too embarrassed to do its best. Your reasons fluctuate, and you may still be unsure about your goals by the end of the day. This is very descriptive of such a mindset.

One of the easiest ways to stop is to take time to meditate. Even 20 minutes of meditation per day slows down your negative patterns of thinking and help you to focus on what is most important. Try it for a few weeks, and remember the habits change over time.

2. see opportunities, not problems.

Writers are a very sensitive group of people, see opportunities, not problems. We also appear to be quite pessimistic, dwelling on, and feeling trapped by the problems. And that can set us up for negative perception and interpretation of all. Whatever happens to us every day, from our reluctance to actually not being able to complete our everyday letters, all these issues are viewed negatively.

Nonetheless, when we move from pessimistic to more positive thinking and consider each problem as a learning opportunity, our cognitive performance improves, and we continue to be much more effective and efficient. Each of our problems includes lessons. All we have to do is take the time to decide what it is.

3. The main thing authors can do themselves is to hope that they will accomplish something that causes them much anxiety, such as sending out manuscripts to publishers. In this way, the author will feel better about sending out her manuscript and imagine how good she feels after the manuscript has really been sent out.

4. Compare negative thoughts to our own worst enemies sometimes.

We should focus on all our negative ideas about what we don't do well. This can create negative thought patterns that can last a long time if we are not careful. It is, therefore, necessary for writers to avoid negative thoughts. You're a new flower and try to cope with a very difficult environment, don't forget. You can do it, but at first, it will be very difficult.

You take really positive action by following these tips to avoid negative thoughts and emotions. It increases innovation and performance overall. And for all the writers, this is a win-win.

We all want to be good writers and happy. We could be hopeful and positive writers one time at a time by taking some constructive steps.

CHAPTER THREE

Anxiety and Overthinking Everything

Anxiety and reversal are bad allies. The tendency to overthrow anything is one of the horrific characteristics is any form of anxiety disorder. Brain Anxiety is hyper vigilant, often looking for something it considers dangerous or worrying. I was fired for making problems in which there are no. Nevertheless, there are indeed problems for me. Why? Because fear causes me to overcome it all. Anxiety causes us to overthink all in many ways, and the result is not helpful at all. Thankfully, depression and overthinking need not be a constant part of our existence.

Why fear triggers overthrow, An impact of any form of panic overturns everything. There are common themes of overthinking depression triggers. Maybe this standard list reminds you of the racing thoughts you experience and makes you realize that you aren't alone in overthinking all because of anxiety.

Obsessing about what we ought / should say /didn't (common in social anxiety) Worrying continuously about who we are and how we measure up to the world (common in social and performance anxieties) Creating fearful scenarios of things which might work wrong for ourselves, the dear ones and the world (common in a general anxiety disorder) Wild, imagined results of our o Wild. Like a gerbil hooked up to an infinite energy drink, they race around in one place and go absolutely zero. The wheel squeaks day and night.

All is a horrific part of anxiety disorders over-thinking. Over-thinking all creates more anxiety. This trick helps to stop worrying further. Take care and overthink it all makes us both tired and wired. One effect of worrying too much about depression is that we often feel physically and emotionally anxious. It takes its toll that these same anxious messages run through all our heads.

However, another dangerous outcome of fear is that we start to think about what we think. After all, it's true if we believe it, and it's very real if we think it's always. Right? Right? No. No. It's tricky anxiety. Anxiety tends to overthink, but these feelings are not always accurate with anxiety.

You have the power and the ability to deal with the overthrow of fear. It is a process that involves several steps, but one action you can take right now to slow down the gerbil has something to do with you or to turn your attention around. Instead of debating or obsessing with your feelings, turn your attention gently to something else, something positive. When worrying about something trivial, you undermine the ability of fear to help you conquer it all.

How to Eliminate Overthinking in Your Life

You might say to yourself, "It is important to me to examine every situation from every angle." But it is not connected to the quality or to the evaluation of a case. You know too much? You think too much? What does it really mean? Every day's experts say that we have 20-42 thoughts every minute. What else are you thinking about so busily? You might think of the mess in the garage, papers on your desk, or even the laundry you need. Some thoughts are insignificant. Whilst others may have more meaning for you than making a telephone call or negotiating a major contract or investing in property in another province or region. You think, and your attitude influences your acts. You think.

You stop going forward and taking action when you spend your time focusing on everything "what if?" What happens when "What if?" isn't happening? You would have been worried about nothing all your time. The most common apprehension is the concept of failure. "What if I fail?" would change to "What if I succeed?" For those who are afraid of failure, fear of success could be just as strong. In the end, procrastination is what it does about thinking.

Concentrate your focus on movement. Sometimes you might feel stuck on the "How can I?" but that's all right. Keep moving, and you're going to create momentum. The "Hows" often appear. Tell others what you want to do and how they told how they did or how they helped in any way to help them. Realize that you will take from them what you want and use it yourself. People are going to ask you how you did it.

Waiting for perfection will rob the world of your talents rather than reaching for all the opportunities available to you. The best way to overcome fear is through practice. Begin by looking at the big picture. What do you want? What do you want? Make a list of all the activities and steps to get there. Perhaps you want them to be classified. Take 3 to 5 moves from your list every day. Before you know it, the big picture begins to form like the pieces of the puzzle.

Stop Over-analyzing Every Thought For Peace Of Mind

A stream of thoughts emerges with a life of its own from your subconscious. One minute it's all right. Next, you're stuck in a damaging net.

The thoughts lead you down a path that overwhelms you.

How does this happen, and why can you get stuck in anxiety?

It is easy to get lost in our feelings because we encounter them tens of thousands of times a day.

Thoughts float across the mind for whatever reason, and they could cause emotional turmoil if we hold with them.

We focus most on our thoughts about our happiness and survival. Situationen that disturb our homeostasis will probably lead to overthinking.

Over-analysis, however, is a vicious cycle that does nothing else than tension.

"As a child, we lose our mental lives... the more we focus on these complex mental events and pay attention, the more complex we are the web of complexity we create," states Master Orgyen Chowang of Meditation, "Our Pristine Mind: A Practical Guide to Unconditional Happiness.

Remember this, when was the last time you thought about an original?

It may have been weeks or months when you last saw one. Because you are used to responding to external events, your perception represents what is actually happening.

Surplus thinking will lead to pressure as our thoughts produce damaging emotions that impact our long-term health.

Our thoughts are harmful if we overanalyze them, rather than let them go unattached.

We are so notorious for recycling thoughts that it defects the present time.

We're not really here, but we're caught up in our heads.

Orgyen Chowang says, "The first step is to understand that your mind is actually uncontaminated, and your mental experiences are just going through. You have to remember that." Remember a time when you have been interested in recreational activities, like a game, an activity, or a time with friends. Note that time has passed, and you are lost in the present moment, not looking to the future.

Being in the flow means being in the field. This means immersing your tasks so that your emotions are real rather than stuck in the past or the future.

In order to avoid over-analyzing thoughts, you must first understand that this is a natural process in which you must work.

To accept that we can not stop negative thoughts means that we are not invested in them. We are not interested in the psychological drama and encourage thoughts to stream unopposedly through the mind.

I appreciate Orgyen Chowang's advice for open-minded meditation. He explains three powerful ways to bring our thoughts back to the present moment through the contemplation of the Pristine Mind: 1. Don't imitate the past. Don't pursue the past.

2. Don't foresee the future.

3. Stay now. Stay at this moment.

And, when we overanalyze feelings, we just attract our consciousness to the habit. This slows down our overactive mind so that we know what is happening.

"We have to recognize that it is not possible to control what comes into our minds if we want peace of mind and more self-control. All that we can do is choose what we believe and do to do," says writer Kelly McGonigal in The Willpower Instinct.

When driven by our instincts alone, we respond to what is happening inside. We feel turbulent thoughts and emotions and wonder how we got there.

What if we have identified that we are overthinking and lean away?

Through following this simple process, the mind is trained to feel thoughts without exhausting you.

Thoughts are like chariot horses, and you're the driver. If you suddenly take off, you can do little to slow down the car. If you take the reins, however, you are well equipped to direct it towards your choice.

It is important to spend time relaxing in silence to understand the essence of your thoughts.

Throughout our modern life, we are filled with sound, and find it difficult to be alone throughout silence.

Nonetheless, we have to set aside our devices, tablets, or TVs to perform other tasks at some point. It means reconnecting the natural flow of the mind with ourselves.

"If you're going to lead a more pacified life, the emphasis will switch as a general practice from outside to internal," Jan Frazier acknowledges, Freedom of Being: Facilitated by What Is. Most people say that they had little time to reflect on, because the lives are far too busy. It is precisely these people who must make meditation a priority.

Just because we can't see our feelings don't all mean well, during a crisis, we can fall into fragments and find peace and happiness difficult to recover. That's because we allowed ourselves to get caught in the stress cycle rather than see it happen.

One good way to stop worrying is to move your body through exercise, even a fast stroll. This harmonizes the mind and body so that we are present, and then live in the past or the future.

Is Thoughts Stress Cause?

Movement involves respiration, which calms the body by activating the nervous system.

We are caught in a sympathetic prevailing state when we overanalyze thoughts. This produces catabolic stress hormones. These stress hormones have a detrimental effect on our health over long periods of time.

The parasympathetic section is the brake of your vehicle, while the accelerator is sympathetic. If you are driving for too long, you will run out of fuel and crash or worse.

The best trend in the western world today is the consciousness that brings our thoughts to the forefront.

Dr. Daniel Siegel, a psychiatrist, and writer, developed a technique to identify destructive thoughts, names and tames, which he calls. If toxic thoughts like fear or anger emerge, we recognize them and call them quietly. In doing so, we become aware of our thinking rather than unconscious.

Otherwise, our minds are at the mercy of dictating our bad moods. This is obvious when the mood changes all day long for no reason. After closer tests, negative thoughts have been swirling and dragging you to a negative emotional state for days.

Psychologists explain how to stop overthinking everything

It's boring to think about something in endless circles.

While everyone worries about a few things, chronic surgeons spend most of their time waking ruminating, putting pressure on them. Then you miss the tension strain.

"There are people who have pathological overthrow levels,"

Pittman is also the writer of "Rewire Your Anxious brain: How to use the psychology of fear in order to relieve anxiety, panic, and worry." Overthink can take a number of forms: deliberating incessantly in decision making (and then challenging the decision), attempting to interpret thoughts, trying to predict future, reading the smallest details.

Those who consistently overthink commentaries, criticize and pick out what they said and did yesterday and fear that they look bad — and are worried about a dreadful future, that' what's if' and' should' prevail over their thinking as if an invisible jury were sitting judging their lives. And they're also nervous about posting online because they are deeply concerned about how others view their comments and notifications.

We don't sleep well because we keep them awake and worried at night. "Ruminators regularly go through incidents and ask major questions: why did that happen? What does it mean? What does it mean? Added Susan Nolen-Hoeksema, Chair of Yale University's Department of Psychology, and Women who think too much: How to Break Free from Reclaiming Your Life." "But they never find answers." The more you constantly focus on ruminating and make it a routine, the more you do it, the easier it is to quit. A clinical psychologist offers a certain perspective. Odessky, the writer of "Stop Anxiety from Stopping You," says, "So often people confuse revision with problem-solving. But what happens is that we just go in a loop," Odessky says. "The overthinking is destructive and mentally draining." We're not really solving the problem. You may feel like you're stuck in one place, and if you don't act, you can have a big impact on your everyday life. It can jeopardize your health and your general well-being. You are more vulnerable to depression and anxiety with rumination.

Many people overthink because they're afraid of the future and what might go wrong. "We are open to the future, and we are always trying to solve problems in our minds," how the brain tricks you into planning for the worst and what can you do about it." It allows us to participate actively in everything around us.

"The occurrence of cardiovascular issues and suppressed immune function is growing from chronic worries. Living in the past or in the future brings us away from the present so that we can not complete the work on our tables. When you ask ruminants how they feel, none will say' happy.' Most of us feel miserable,' said Nicholas Petrie, a senior lecturer at the Creative Leadership Centre.

Thinking over can trap the brain in a loop of concern. If ruminating is as natural as breathing, you must deal with it quickly and come up with a solution.

"If an unpleasant event puts us in a disgusting mood, it is easier to remember other times when we feel awful. This can put a ruminator on the stage to work into a downward spiral, "Amy Maclin of Real Simple wrote.

How to change this habit of thought and win your life is not easy. Constant anxiety. It is a pattern of thinking which can be broken. You will train your brain to take a different view of life.

Pittman recommends that you substitute your thoughts in order to overcome overthinking. "Telling yourself not to think is not the way not to think," she says, "you must substitute the thought." What if she should tell you to stop thinking about rose elephants? What will you think about it? Okay: pink elephants. If you don't want an elephant blue, conjure up a tortoise image, say. "Perhaps a big tortoise has a rose in its mouth as it crawls," Pittman says. "Now, you don't talk about pink elephants." Talk out of it when you find that you are stuck in your head. You will master your overall habit if you can start your self-talk— the inner voice which gives a running monolog all day and into the night.

"With other interpretation of situation, you can cultivate a slightly psychological distance, that allows the negative thoughts less trustworthy. This is referred to as mental rehabilitation.

Ask yourself—What is the likelihood that what I'm afraid will happen? What are some more likely outcomes if the likelihood is low?

If you continue to ruminate about a problem, rephrase the matter to reflect the positive outcome that you are looking for, "says Nolen-Hoeksema.

"Responding to yourself, or even better, writing, instead of" I'm stuck in my career," "I want a job where I feel more committed. "Then plan to expand your skills, network, and find opportunities for a better career.

Consider a positive way to treat any doubts or negative thoughts, Honey says. "Every night before bed or the first thing of the morning, write down your thoughts in a diary— they don't have to be in any order. Make a' brain dump' on the website of everything on your mind. This can sometimes give a sense of relief.

By connecting with your senses, you can also control your ruminant habits. Start to see what you can hear, see, smell, taste, and feel.

The goal is to reconnect to your immediate world and all around you. You spend less time in your head when you start to notice.

You can also notice and talk about your overthinking habits. This will help you gain power by becoming self-aware.

"Wait a bit more," Carbonell says. "Say something like: I feel somewhat anxious and uneasy. Where am I? Where am I? Am I in my head? Am I in my head? Perhaps I should walk around this block and see what is happening. "Recognize that your brain is overdriving or ruminating, and then try to snap out instantly. Or rather, distract yourself and turn your attention to something else that requires concentration.

"If you have to pause and fix hundreds of times a day, it will stop quickly, perhaps within one day," says Dr. Margaret

Weherenberg, a psychiatrist who has written The 10 Best Ever Stress Management Techniques. "Even if the move is simply to return focus to the job, a decision must be made to change ruminant thought." This takes concrete effort, but over time you can easily understand when you stress unnecessarily and instead choose to do something in real life instead of spending lots of time in your head.

"I can't believe it happened," for example, transform to "What can I do to prevent it from happening again?"Or convert," I've got no good friends!"To' What steps can I take to strengthen my friendships and find new friendships?. Suggests. Don't get lost in wondering what you might have, what you might have, and should have done differently. Mental stress can have a serious effect on your quality of life.

An overactive mind can miserable life. Learning how to stop taking time in your head is one of your best gifts.

Changing your destructive patterns of thought, like all habits, can be a challenge, but it is not impossible. You could Improve your life to see things in different ways and reduce them. overthinking stress.

When rethinking is ruining your life, and you fear that because of your thoughts, you may be spiraling into depression, professional help is worthwhile.

CHAPTER FOUR

Dealing With Emotional Stress

Emotional stress is a very difficult form of stress to cope with and manage. After all, it is often self-created, may come from nowhere, and the stress caused just increases the feelings. Therefore, as emotional stress decreases, emotions get worse, and emotional stress is increased. The problem thus recreates the cause and only worsens the problem.

Emotional stress is often caused by a dramatic occurrence, which makes the nervous system of a person under severe strain. This could be an event like the loss of a dear one, the death of a person, or a life-threatening situation. An occurrence such as this can make a person's mind, and nerves extremely stressed, and extreme stress can change the way the brain works. Indeed, a severe emotional strain might even lead to post-traumatic stress disorder.

Emotional stress, however, is not caused by a sudden shock. It can also result from an unbearable pressure that prevents someone from talking about something other than the issues that seem to be unresolved. Instead, as tension increases, the mind is trapped in its own stress cocoon, which can only draw attention to itself and cut people off from the outside world. Therefore, emotional stress can cause isolation and unable to focus, tiredness, and even memory problems.

Unfortunately, emotional stress often raises moods, often exacerbating issues. In reality, emotional excess attacks can lead to excessive emotional excess, leading to further attacks. Instead, as these emotional stresses continue to add up, all becomes too much, and the patient remains almost totally and alone in the emotional cycle that constantly pounded into the brain.

The person who suffers from this must take a break from everything that causes all emotions in order to cope with emotional stress. For example, taking a holiday can be great fun as it gives the brain new inputs that do not require associations. When a person suffering from emotional stress leaves the so-called "scene of the crime," he/she will remove some emotional stress by removing it. Then the loop will hopefully be broken so that the person can start fresh.

Another effective way to deal with emotional stress is through yoga or meditation. Exercises like these are designed to put the mind of a person into the moment so that they don't care about anything but what they do for yoga or, in the case of meditation, completely clear the mind and encourage him to throw away his emotions and begin again with a slate. Both methods can be very efficient to deal with emotional stress since they give the brain an opportunity to relax. Then, once it is relaxed, it can relieve emotional stress and return clearly to the business of thought.

Similarly, a hobby can be great for emotional stress. An event like needlepoint, model building, sporting, or going fishing may lead to stress reduction. This is because a person in a hobby enjoys himself while thinking just about what he is doing rather than everything that has to be done. It is like a cross between a vacation and relaxation, where the person takes a break from life and focuses the mind on something else. The tension, therefore, goes away, and the person may feel that he is doing something, even if it's only a small, insignificant achievement. After all, a worthless success remains. It's always nice to know that some work is successful, and hobbies are a great way to become productive unexpectedly.

Emotional stress should not intimidate people. They should instead try to understand where it comes from and what they can do to prevent it. Even if the effort can often be challenging, success is indeed its own reward. After all, getting away from emotional stress provides the mind with instant benefits and long-term benefits for the body. In addition, people can see what causes emotional strain by understanding it and hopefully discover what they have to do to either cope with it or completely eliminate it. So, if you or someone you know has emotional stress, find solutions that work. And if emotional stress continues to be implemented, it can become a thing of the past. So, if you or someone you know has emotional stress, find solutions that work. And if emotional stress continues to be implemented, it can become a thing of the past.

Stress and Psychosomatic Disorders

It has long been known that negative emotions are linked to specific diseases, such as fears leading to cardiovascular diseases, liver damage caused by anger, apathy, and stomach disorders. They all share something-stress. Stress. But how are we supposed to deal with stress?

What is stress?

Why is it happening?

Is it always bad? Is it always bad?

Stress is an unavoidable part of daily life. Lower stresses are harmless (and sometimes even helpful), but negative, long-lasting stress can weaken your health.

The pressure was defined as a series of normal genetically programmed, non-specific reactions of an organism to its survival through the "fight or flight." It was the founder of the concept of pressure. Minor negative effects usually do not cause stress. It happens when the stressors exceed our ability to deal with them. The stressors change the functioning of the body by mobilizing its resources to deal with danger (increase blood pumping and expansion of the airways to increase the intake of oxygen, increase blood coagulation, etc.) or adapt to it. This is the main aim of the stress reaction.

There are three phases to typical stress response: alertness—to mobilize any protective means in the body.

Stabilization-balanced use of the adaptive ability of the body.

Exhaustion—The final phase after the cumulative impact of stressors has depleted the body's adaptive resources.

Stress is a natural part of life, producing a "life flavor" in Selye's view. Stress helps us in dynamic work processes, creative activities, and competition. while the factors have a strong influence is overblown and continuous, they drain our means of protection and may cause disease or even psychosomatic or neurotic disorders.

Different people react differently to stressors. Some people react proactively to the threat. Others respond passively and abandon quickly. These types of reactions generally cause certain types of disorders. Based on several medical findings, doctors found that most stressors usually cause high blood pressure, ulceration, heart attack, stroke, heart rhythm, etc. Angry that is not expressed can cause rheumatoid arthritis, skin problems, migraines, indigestion, etc. When we feel strong negative emotions, there are important physical changes in the body that cause excessive energy production. In addition, a chronic negative psychological position/personality also causes a rapid depletion of the protective measures of the body.

Combination of stress and illness.

Psychiatrists and Psychologists found strong links between certain personality characteristics at one end and somatic disorders of the other. Example: people trying to fit into a certain position/job that does not match their personality or abilities have a greater chance of developing cardiovascular disease. Chronic coronary disease is more typical for proactive, ambitious, and less tolerant people.

Persons with stomach ulcers are usually very nervous and irritable. They are quite conscientious, but they typically have low emotional-esteem, are susceptible, shy, hypochondria, and sensitive. Some people always try to do more than they can. We prefer to overcome problems with very high anxiety.

The level of psychological changes associated with depression in the body is generally linked to a personal assessment of the situation, which in turn depends on feelings of personal responsibility. Signs of emotional tension in stressful situations tend to intensify when physical activity is lacking.

Disorders with stress.

There is no clear list of disorders associated with stress. The same type of disease could be caused by stress or something else. Some variables in a person's life can be associated with the pressure that has a negative influence on the function of the body. The combination of negative factors is particularly dangerous because it creates more opportunities for certain diseases to develop.

The neuroses-mental disorder caused by protracted psycho-emotional expertise, mental and physical strain, lack of rest or sleep, long-term internal struggle, inhibited the sense of sorrow, anger, or suffering-belong to the many stress-related disorders. Many somatic disorders can also induce neurosis.

Neurosis may occur because of a lack of solutions to a serious problem. It can happen if a person tries to solve a problem, but can not. This results in increased vulnerability or irritability to the problem that makes a person emotional. This causes the individual to experience different pains in response to stressors.

The reactions to stress are very different.

A satisfactory emotional benefit increases the quality of an individual. Nevertheless, excessive emotional stress leads to a decline in performance. The more complicated the operation, the more pain the person gets, causing tiredness, apathy, attention loss, fatigue, and memory difficulties.

Some people may very actively respond to stress, while others will quickly give up. A constructive response can lead to snap decisions that only concentrate on the key aspects of the problem. The hyperactive-compulsive reaction led to a considerable increase in error, while activity remains strong or even increased. In contrast, an inhibited reaction leads to slower thinking and increases progress in the learning process.

The psychological working and home environment plays an important role in maintaining physical and mental health. Everyone's mood depends a lot on the moods of the people around him, and in his gestures, imitations, and actions, it shows up. When you communicate with other people, you tend to be optimistic or depressed. And shared solidarity is a common indication that the team of co-workers or family members has a good atmosphere.

No one is free from injuries, failures, or problems that can not be resolved. Nonetheless, it's not good to focus too long on negative emotions or to overwhelm you with depression. It is much safer for your wellbeing to focus on trying to find a positive solution.

How to alleviate stress.

Exercises isometric. This approach is based on the pressure exerted on some muscles and then released rhythmically. For example, make a fist and then relax it, or put your hands behind your head, and force your muscles to pull back, then push your feet to the ground and relax. Such basic exercises can be used to relax in any situation.

Self-engineering practice. This is a well-known relaxation method. A deep relaxation that people usually experience after hypnosis can be accomplished by self-suggestive techniques. You can sit still and make simple orders like I feel calm, I feel heavy, my arms and legs warm and heavy. To do this really well, you have to practice it a few times. This technique can have a stronger effect if deep respiration is used during the exercise.

Therapy. Meditation. All known methods of meditation are designed to focus your thoughts and attention on one thing. It could be music, chant, or breathing for yourself. All other thoughts are shut down, and other distractions are totally ignored. This concentration allows you to relax fully. Deep breathing with a certain sitting posture and closed eyes helps to get complete rest.

Training on biofeedback. Biofeedback has become very common among stress management practitioners in recent decades. The concept is based on the calculation of certain physical parameters, which can be regulated directly or indirectly. The subject/person can see the actual levels by offering different suggestive or control commands that influence the calculated parameter. One example is the continuous measurement of skin temperature, while suggestive thoughts are induced to muscle relaxation. Muscle relaxation allows the peripheral blood vessels to dilate and to increase the flow of blood in the body. The temperature rate is shown to the subject, and suggestive thoughts are continued. The topic thus provides feedback that contributes to deeper relaxation.

A special biofeedback approach has received considerable attention in the area of stress management over the past decade. It is based on the use of deep rhythmic breathing, which positively influences the rhythm of the heart by causing it to oscillate regularly. This approach incorporates a very critical Baroreflex physiological mechanism which is responsible for adapting the body to various factors (physical exercise, psycho-emotional stresses) and maintaining internal homeostasis. The method is special training to this system by exercising and toning it in a similar manner to physical muscle and cardiovascular exercise.

During the workout, your heart rate is measured and shown. At the same time, the trainee requires a visual and/or audio pacer to maintain a certain breathing pattern. A special mathematical algorithm evaluates and continually shows the effect of paced breathing on the rhythm of the heart. This algorithm analyzes an accuracy rate of around six respirations per minute between your heart rhythm and your breathing rhythm.

The immediate effect of such training is stress relief and restores the internal equilibrium of the body. Regular use of the technique causes several positive effects such as reduction of blood pressure, strengthening of the immune system, improvement of digestion, normalization of metabolism, and chemical balance.

Benefits of a Decluttered Mind

Advantages of a Decluttered Mind

The best spot to start to declutter your life is from within. Numerous individuals neglect the advantages of a sound mind can offer. The mind can move toward becoming hindered with psychological weight and indeed sway an individual's capacity to work. Necessary leadership can turn into a test and adapting to issues may feel about unthinkable when you don't have a clear mental state; in this manner, it is imperative to figure out how to free your mind of excessive clutter. Since everybody is different there is nobody size-fits-all strategy to clear your mind of clutter; nonetheless, coming up next are some basic methods that can start you on your adventure to decluttering your life by first decluttering your mind!

All in Good Time: Schedules

Lighten a portion of your mental stress by making a timetable. When you have all assignments sorted out and arranged out, with spare time included between, a significant measure of strain will be lifted. You will live more proficiently and suffer from less overpowering minutes. You can't get ready for everything, and calendars must be changed now and again. In any case, having a reliable schedule for the things you realize you should do, organized by significance, can have a considerable effect on your mental stress. Besides, this is an incredible method for promising you to have time put aside to rehearse your mind-decluttering systems.

Meditation for a Clear Mind

Meditation is a famous instrument to help declutter your life and your mind. You don't need to think as it was done in the good 'old days. Attempt this increasingly modernized system: start with music you appreciate. A few people profit by uplifting tunes or great songs, while others may prefer something edgier. The class is altogether up to you, and it doesn't need to be unwinding music. Next, locate a single region that you can disengage yourself from others and diversions.

Start by playing music. It is useful to have a playlist, or you can circle a similar song if you prefer, as long as you don't need to get up and restart the music when the song closes. The music is going about as a manual for the assistance you start to declutter your life. Get settled on a bed, love seat, seat or the floor. If you rest, ensure you don't nod off! Tune in to the music as it fills the zone around you. Close your eyes and spotlight on it. Enable your body to unwind. It is ideal for lying on your back or sitting with your arms limp to advance total unwinding. Choose an essential sound in the song and tail it. Give your mind a chance to move toward becoming immersed in it as you trail the sound through the song. This enables your mind and body to concentrate on something different while unwinding and getting a charge out of good music.

Use Words to Remove Tension

Written words are an integral asset to declutter your life. How you utilize them is up to you. A few people prefer to write in a diary. This can be private, and nobody else needs to see it. If you are worried about others discovering your written musings, consider writing them down on a paper and after that discarding it or demolishing it after you are finished.

Another suitable method to utilize written words is to compose letters. This is frequently done when pessimistic feelings emerge towards someone else in your life. When the message is written, store it someplace or discard it. The thought is to get your feelings out on the paper, instead of on the individual. This can work another route too. When you are feeling discouraged or down, pen a positive letter to a companion. This will help remind you of the good things so you can stay focused. You can even send the message if you need to! It is imperative to recall when you declutter your life, and you should endeavor to expel negative feelings and stay focused on the right things.

Start to declutter your life presently, beginning with your mind. You will feel better, work all the more adequately, and suffer from fewer misfortunes. Besides, when terrible things occur, you will be better prepared to deal with them when your mind is clear of clutter!

It's Time To Declutter And Organise

Why is cleaning up so difficult and anguishing? Getting rid of the gear that you have claimed for a very long time can generally make some nervousness and tenseness in your mind, frequently emerging from an extent of clashing considerations and feelings.

"Should I toss it or keep it?" This is maybe one of the most baffling inquiries that numerous householders face regularly. The fact remains that there are just a couple of anxieties that should be identified and dealt with, yet cleaning up turns out to be such an agonizingly overwhelming undertaking.

One of the most well-known approvals that we give ourselves is, "Why toss it? It'll come to utilize sometime in the not so distant future", and on this guise of we continue collecting new clothes, furniture, and other household stuff. "It had cost me a bomb!" "Goodness! That is familial!" "That was a gift! So inconsiderate of me to relinquish it!", and the reasons start taking the better of us. So the minute the thought manifests itself, that you have to clean up, 'Simply let it all out!'

You have to remind yourself about a bunch of things. One that you can generally resale your costly, unused assets and somewhat conceal the cost you paid to claim them. Like this, it doesn't consume a significant gap in your pocket. For instance, that old unused China enhancing your porcelain bureau for a very long time should most likely be possible away with and will draw in a lot of purchasers as well. Two, what is repetitive and not required by you, maybe extremely valuable to your companions. They may be satisfied at your motion of giving, and you can relax in the passing praise you get, rest guaranteed that your things will be utilized and dealt with. Take, for example, that creator T-shirt that you battle to fit into or that costly pair of thin pants that are merely lingering its time away in your cabinet; someone can unquestionably utilize these better. Three, you have almost every time pulled through the lost, misplaced or spoilt stuff, even the chapter s that were important like that ravishing table top glass that is broken now or notwithstanding something like your misplaced mobile phone. Henceforth, it is only this one snapshot of shortcoming that you don't need to give in. Let yourself know that you can live without it, and it is only a worldly belonging.

When you are finished arranging off the useless stuff, the time has come to organize and orchestrate what remains. You can, in fact, productively organize your things in the much unthought-of places. Here are a couple of arrangements that can genuinely prove to be useful.

Taking off drawers fitted under your beds can without much of a stretch stack up against your bedsheets, covers, and even clothes. An adroitly arranged heap of drawers can substitute a regular staircase if there should be an occurrence of cots, and all the kiddo stuff like toys, clothes and even books can be taken care of correctly in them. Pressing sheets nowadays can be taken care of in haul out drawers, all you need is keenly planned cabinetry. A staircase can end up being an impressive extra room with a few risers transforming into haul out drawers where you can reserve up shoes, cleaning supplies, and so on. Fitting in an electrical extension board at the backside of a drawer can give your gadgets a chance to energize inconspicuously. The space underneath the kitchen sink which we frequently leave as unusable can, in fact, be fitted in with equipment to suit your sink pipe, or rather shroud it and have room for all your cleaning supplies. Furthermore, the rundown goes on.

Advantages of cleaning up and getting sorted out are various, both as far as physical and mental. Rest will be dropped by simple and tranquil once the house is clean and free of messiness. You will feel more empowered and will have the option to deal with life better. Because you will invest less energy cleaning, you will have the opportunity to set aside out effort for your side interests and premiums and will have a chance to ingrain in your youngsters the great propensities for cleanliness and association. The stuff you give away will without a doubt, come being used to somebody and the best part is that your whole home will look prettier and shimmering clean.

When it comes to cleaning up and arranging your home, you need to understand that it didn't turn out to be unorganized from only one day disregard. In this manner, you can't anticipate that it should just require one day of exertion for it to move toward becoming de-jumbled. It's useful if you can plan yourself a specific measure of time regularly to pick away at this issue which will make what appears to be an enormous endeavor significantly increasingly sensible. Preceding beginning helps yourself out and snatch yourself three boxes and set them up this way. Mark one to "set away," one to "give away" lastly one to "discard." You might need to line your "cast off" box with a junk pack so it will be simpler to discard the items later too.

Your objective to getting to be organized is most effectively accomplished by cleaning in an organized way that way you'll focus on only each room of your house in turn. Start where you go into the room and clean a clockwise direction around the room, be tenacious, and don't merely bypass zones. You'll most likely think that it is difficult now and again to choose which items are essential and that ones that you genuinely need to keep, and after that which things ought to go in the "give away" or "discard" boxes. This is typically the most testing part, so how about we talk about how to deal with that.

To help settle on your choice, ask yourself a couple of inquiries about the item to decide how you feel about it.

1.Do you truly adore it?

2.Does it mean something to you, or does it have some degree of sentimental worth?

3.Does it make you at all feel guilty or miserable for a specific reason?

4.Also, have you utilized the item in the previous year?

5.Do you have a comparable item that might be of better quality?

6.Is it broken?

These are mostly quality reason not to keep a specific item. Remember because somebody gave you an issue and you feel guilty disposing of it isn't a sufficient reason to keep it. To utilize this procedure as your outline and manual for settling on choices on what part with.

When you have filled you "give away" put away, you can either list the items on Craigslist or eBay or permanently drop the case at a gift focus or thrift store.

It is substantially more apparent once you start to de-mess one room and look in your "set away" box, exactly what number of items are in reality, strange in your home. When this case is full, deal with it right away. Keep away from the allurement of reserving it away in the storm cellar or storage room. You may plan on taking care of it later; however, chances are you presumably won't.

While you are dealing with your assigned room, be mindful to not skirt drawers, racks or furniture pieces, for example, a work area. Organize each drawer and designate a place for items, for example, charges stamps, pens, and envelopes. If this is the place you store money related or other important papers, experience them also. Anything over a year old that you have to keep, consider putting away in a safe and ideally flame-resistant elective place for a long time.

It is substantially more apparent once you start to de-mess one room and look in your "set away" box, exactly what number of items are in reality, strange in your home. When this case is full, deal with it right away. Keep away from the allurement of reserving it away in the storm cellar or storage room. You may plan on taking care of it later; however, chances are you presumably won't.

While you are dealing with your assigned room, be mindful to not skirt drawers, racks or furniture pieces, for example, a work area. Organize each drawer and designate a place for items, for example, charges stamps, pens, and envelopes. If this is the place you store money related or other important papers, experience them also. Anything over a year old that you have to keep, consider putting away in a safe and ideally flame-resistant elective place for a long time.

De-jumbling your home may feel extremely overpowering from the start however it isn't generally as awful as it appears. You truly need to focus on following your work routine and take as much time as is required. It is likewise useful to keep a cleaning cloth with you while you're sorting out so you can wipe things down too. After you are completed de-jumbling the room you are dealing with, feel free to enhance your room with any exceptional items to make the finished outcome precisely how you need it to look.

CHAPTER FIVE

Take Time to Declutter Your Life

When you think about clutter, do you discover stuff or turmoil around your home? That is entirely one importance of clutter, yet clutter is a long way past that. Clutter can be inadequate exercises that are hanging over your head or tolerations that you suffer in life. Clutter is undesirable relationships, grievous inclinations, negative considerations, and hazardous energetic states. Clutter is whatever interferes with carrying on with your best experience and being your best self.

Why care about the clutter? Clutter is a minute essentialness channel. Notwithstanding whether it's jumbled storerooms, a terrible attitude, work you couldn't care less for, or unpaid charges, they all interfere with carrying on with an upbeat and fulfilling life. When you discard the clutter, you adore yourself enough to develop and be your best. You know the well-known proverb, "Time to proceed onward to better things." Address clutter in your life opens the door for new and invigorating things to happen. Each time I say goodbye to something that is never again agreed with my best involvement, another open entryway strangely appears.

There are four critical aspects of our lives that clutter will, as a rule, gather.

Physical Environment

Your physical condition involves your home, office, and vehicle. Decluttering your surroundings joins throwing out the trash and organizing what you wish to keep. When you walk around your space, OK state you are upheld by it? Is it immaculate and lit up in a way that brings you concordance and fulfillment? Decluttering your condition suggests that everything in it is in the high working solicitation. Right now, my vehicle needs a couple of fixes, and it impacts me. My atmosphere control framework doesn't work. The side of the driver's window doesn't open, the electric locks make an uproarious squashing confusion, and the fuel injector siphon is going out. Does my vehicle run and get me around? It does, yet it's aggravating to have these outstanding subtleties not managed. I detest driving my truck. How is your environment? Do they enable you to live relentlessly and merrily?

Prosperity and Emotional Balance

Your physical and excited prosperity is about you. Decluttering yourself physically and truly infers you are figuring out how to eat right, practice reliably, and manage your mind and body. You are tending to any prosperity concerns and keeping standard checks ups with your PCP. Decluttering force you to address stress in your life and find a way to live a quiet nearness. Keeping up energetic leveling consolidates managing your idea and eager experience - doing your best to keep your mind positive. It suggests you are rationally fortifying your brain, empowering the creative mind, and keeping up a vital good ways from the things that are hurting to your mental state, for example, obsessive worker conduct or watching a ton of TVs.

Cash

Decluttering in the territory of cash anticipates that us should keep up significant budgetary affinities. What is a sound cash chief? A vigorous cash supervisor has watched out for all her essential topics with cash. She is available to deal with cash and uses her cash cautiously. Decluttering anticipates that you should address overspending or living in haziness when it goes to your budgetary picture. Having substantial money related inclinations infers you are saving money for both present minute and whole deal needs. You have a will that keeps an eye on all of your preferences, including your children. Individuals in a sound budgetary state grasp the advantage of giving and not additionally holding solidly to cash. Bills are paid on schedule, and commitment is non-existent, aside from a home advance. A reliable cash administrator is told in smart endeavors or has a reliable and trusted in budgetary insight. Cash is a contraption that is indispensable for you to live in a way that is basic to you.

Relationships

Relationships in your life join your family, allies, associates, and boss. Decluttering in the zone of relationships suggests you are tending to any relationship issues and that you, all around, coincide well with the individuals in your life. You have ousted the associations from your life that reliably drag you down or hurt you. Keeping healthy relationships anticipates that we should remain in contact with individuals by eye to eye, phone, or email visits. To declutter your relationships, you need to acquit everyone who has hurt you and put a full end to the relationships that are no longer in your life. Doing your part to keep up healthy relationships anticipates that you should talk honestly, refuse to snitch, and abstain from examining and settling on a choice about others. When we've decluttered our relationships, they give the adoration and support we need on this experience through life.

Explore your life. Where are you have the option to use a touch of decluttering? What necessities to change to empower you to carry on with your best experience and be your best self? Our priest shared an idea that benefits were holding tight to: "If I need my life to transform, I should transform." I'll make it one walk further. "If I have to develop and carry on with my best life, I ought to develop and be my best self." refuse to snitch, and abstain from examining and settling on a choice about others. When we've decluttered our relationships, they give the adoration and support we need on this experience through life.

Explore your life. Where are you have the option to use a touch of decluttering? What necessities to change to empower you to carry on with your best experience and be your best self? Our priest shared an idea that benefits were holding tight to: "If I need my life to transform, I should transform." I will make it one walk further. "If I have to develop and carry on with my best life, I ought to develop and be my best self."

Countering Overthinking: A Step Towards an Improved Life

So let us say that you're hanging around at a meeting with colleagues and customers, and you've found someone you really want to talk to. Perhaps it is related to business, or you only want to establish personal ties. Whatever it is, you start preparing a psychological draft of what you have to say and plan on meeting them, but a shaking fear in your head stops you on your journey. What if they don't want to talk to you? What if the conversation line doesn't work? Or is it going awfully wrong? Your fear creates a kind of domino effect, and you start thinking of the worst that could be inevitable. Each thought draws you deeper into the tangled mess of confusion within your mind, making you ultimately unable to talk even more. You then watch another person talk to the subject: a missed opportunity.

Surprisingly enough, uncertainty and consequent restlessness and anxiety, while proving to be a massive dissuasion in one's social and personal lives, also become the trigger of lost opportunities and moments that one will later apologize for. But it could be easily overcome with a few daily practices and a certain attitude.

Accepting The first step toward anxiety and overthinking is first and foremost to accept the problem. Only then can you go ahead and fix it. But even though you know that you are a remunerator, it is also important to realize that you are not alone and that there is no reason to be afraid. Surplus thinking is a normal thing for many nowadays, and you could resolve it with a positive attitude.

The best moment is now. The best thing you can do without overthinking is, of course, to stick with the present. Your brain can not think of things far away if you're busy where the flow should be. You also learn to appreciate your environment and presently significantly improving your performance on any task. And while it is much easier to say than do, there are some strategies you can use every day to reduce to a large extent the cycle of negative thoughts.

Breathing is, indeed, a good example. How much that benefits, you would be shocked. Just close the eyes and take a few minutes to breathe deeply. Watching carefully and breathing deeply allows you to take in the current moment and helps clear your head.

Another good example is to meditate for awareness. The basic idea is to remain silent and just to concentrate closely on everything around you, which has done wonderful things for many. Just once a day, close your eyes and try to take in your entire environment. Listen to your thoughts but don't' interact' with them, and you can actually try to change their' size.'

Therefore, slow down. Do all that you do with your full awareness. Try and tell each step you take to yourself and force yourself to notice your environment. This will also motivate you to stay right now.

Trust When you are full of self-esteem and feel good about yourself, you create a positive attitude. You would be less likely to overthink, so anything you say or do turns out to be done better. One of the first things you can do is get involved. Create a schedule of what to do and continue to be successful for the day. Doing things prevent you from wandering away, and in addition, getting things done results in a strong boost in confidence by a sense of achievement. You should also try, and at least once a day does something you really are good at. Whether you are an instrument player, or have a tremendous talent for a video game, take some time off your schedule and do it. It's going to be a big help.

One life that changes you can make is to counterfeit it. It could sound hard, but it really works very well. Pretend that you're a person you know who's witty, intelligent, and self-assured. You might recognize one from a TV show, a movie, or a novel. Go on and deliver with faith all you say, even if you're not sure, or you're terrified. You will find that as you counterfeit it more and more, you will finally bear the confidence in real life.

Let's go and try to control all the outcomes of your life is certainly the main cause of pay. Because if you do, you are also doomed to think feverishly about what to do in every moment of your life, wondering what could happen next. The best thing you could do is not to convince yourself. Realize that in what takes place in your life, you have no say, and there is no reason to worry about it. The world has determined your fate, so you should make the most of every moment. Try to realize this before you can hesitate to do anything, and it will help you stop thinking over and just do it.

You can also make certain schedules to make a decision. Whether it is going to talk to someone or to make wider choices in life that may force you to overthink, take one minute for the little ones, and a few days for the bigger ones in life. This would encourage you to evaluate a decision and make the best choice rationally. Steel yourself and do it once you make a decision. It may be terrifying, but in the end, you will find it satisfying.

So let us say that you're hanging around at a meeting with colleagues and customers, and you've found someone you really want to talk to. Perhaps it is related to business, or you only want to establish personal ties. Whatever it is, you start preparing a psychological draft of what you have to say and plan on meeting them, but a shaking fear in your head stops you on your journey. What if they don't want to talk to you? What if the conversation line doesn't work? Or is it going awfully wrong? Your fear creates a kind of domino effect, and you start thinking of the worst that could be inevitable. Each thought draws you deeper into the tangled mess of confusion within your mind, making you ultimately unable to talk even more. You then watch another person talk to the subject: a missed opportunity.

Surprisingly enough, uncertainty and consequent restlessness and anxiety, while proving to be a massive dissuasion in one's social and personal lives, also become the trigger of lost opportunities and moments that one will later apologize for. But it could be easily overcome with a few daily practices and a certain attitude.

Accepting The first step toward anxiety and overthinking is first and foremost to accept the problem. Only then can you go ahead and fix it. But even though you know that you are a remunerator, it is also important to realize that you are not alone and that there is no reason to be afraid. Surplus thinking is a normal thing for many nowadays, and you could resolve it with a positive attitude.

The best moment is now. The best thing you can do without overthinking is, of course, to stick with the present. Your brain can not think of things far away if you're busy where the flow should be. You also learn to appreciate your environment and presently significantly improving your performance on any task. And while it is much easier to say than do, there are some strategies you can use every day to reduce to a large extent the cycle of negative thoughts.

Breathing is, indeed, a good example. How much that benefits, you would be shocked. Just close the eyes and take a few minutes to breathe deeply. Watching carefully and breathing deeply allows you to take in the current moment and helps clear your head.

Another good example is to meditate for awareness. The basic idea is to remain silent and just to concentrate closely on everything around you, which has done wonderful things for many. Just once a day, close your eyes and try to take in your entire environment. Listen to your thoughts but don't' interact' with them, and you can actually try to change their' size.'

Therefore, slow down. Do all that you do with your full awareness. Try and tell each step you take to yourself and force yourself to notice your environment. This will also motivate you to stay right now.

Trust When you are full of self-esteem and feel good about yourself, you create a positive attitude. You would be less likely to overthink, so anything you say or do turns out to be done better. One of the first things you can do is get involved. Create a schedule of what to do and continue to be successful for the day. Doing things prevent you from wandering away, and in addition, getting things done results in a strong boost in confidence by a sense of achievement. You should also try, and at least once a day does something you really are good at. Whether you are an instrument player, or have a tremendous talent for a video game, take some time off your schedule and do it. It's going to be a big help.

One life that changes you can make is to counterfeit it. It could sound hard, but it really works very well. Pretend that you're a person you know who's witty, intelligent, and self-assured. You might recognize one from a TV show, a movie, or a novel. Go on and deliver with faith all you say, even if you're not sure, or you're terrified. You will find that as you counterfeit it more and more, you will finally bear the confidence in real life.

Let's go and try to control all the outcomes of your life is certainly the main cause of pay. Because if you do, you are also doomed to think feverishly about what to do in every moment of your life, wondering what could happen next. The best thing you could do is not to convince yourself. Realize that in what takes place in your life, you have no say, and there is no reason to worry about it. The world has determined your fate, so you should make the most of every moment. Try to realize this before you can hesitate to do anything, and it will help you stop thinking over and just do it.

You can also make certain schedules to make a decision. Whether it is going to talk to someone or to make wider choices in life that may force you to overthink, take one minute for the little ones, and a few days for the bigger ones in life. This would encourage you to evaluate a decision and make the best choice rationally. Steel yourself and do it once you make a decision. It may be terrifying, but in the end, you will find it satisfying.

Understand What Happens When You Are Emotionally Stressed

Many people find that they are exhausted emotionally and physically when they live in a stressful or uncomfortable situation. Everything can not be seen in a calm and clear manner. Life is becoming difficult; relationships are becoming fragile.

Take time to explain what occurs when you're emotionally anxious-recognize that you live in an emotionally challenging situation. When you have an overly complicated emotional life, then you might need to speak with your partner to discuss your emotional state. It may be time to decide whether the relationship is worth the emotional trauma and discomfort, can it be handled better, is it worth to you?

You might have to consider what signals you send to others? You may come across as unfriendly, disinterested, even hostile when you are stressed and tense. If so, then it is understandable that your relationship is out of control. Your partner should allow you a wider berth than normal, so you can relax and regain your balance.

When you feel stressed, it can lead to your partner, not listening, and taking care of what is happening in his / her life. In order to address the issue and strengthen the relationship, you may have to realize that you have become so concerned about your own concerns that you do not have time or space to take account of the needs of others. Stress can make some people so miserable that they immediately begin to believe that everything wrong is their fault.

If you are depressed, it is important to remember that not everything is about you in life. You can feel that others judge you, avoid you or even ignore you, but there are several explanations for this. Did you misread their signals? We often take everything personally when we feel stressed. People smiling, looking to us as they speak can all be perceived in a negative way.

Be open to the explanations for the actions of others. We can also be depressed or over tension in their lives. You may not want or feel willing to talk to you about your distress. You may not be able to appear insecure to you or may not be confident in helping you.

If we are emotionally depressed, negative emotions will start to emerge:-Jealousy can become part of our mindset as we ask what the other person wants or does. You can begin to be totally out of character, check the messages and numbers of your partner's phone, check their pockets for evidence of another life.

Anger and resentment will arise when we start to blame others for making us feel like this. Begin to take care of your feelings. Nobody can make you feel anything unless you invite it.

The effect on your sleep, libido, laughter, health, consuming alcohol, eating habits, energy levels can be a result of the long-term negative emotions in your life. General health and well-being can all be compromised if you are depressed and mentally stressed.

Hypnotherapy might play a useful role in coping with those negative emotions, knowing the underlying reasons for your actions, figure out why it was caused, and recover a better sense of balance.

Techniques to cope with and communicate effectively with stress, understanding the importance of engaging in regular breaks and caring for oneself are all vital for relaxation, stress management, and accountability for mental health and happiness.

CHAPTER SIX

Using Aromatherapy to Support Symptoms of Depression

A fantastic and mysterious relationship between plants and their environment is continually unfolding. These organic green machines perform glorious alchemy regularly with water, soil, air, and sunlight. The almost infinite combinations of plant genetics and the environmental conditions that surround the earth have provided for a virtually unprecedented variety of alchemical floral expression and resulted in a wide range of natural botanical materials. This includes simple essential foods, vegetables, and gourmet fruits, useful medicinal herbs, rich exotic spices, sophisticated and enchanting natural perfumes, crucial therapeutic oils. Humanity reaches into the jungles and rain forests evermore, knowing that nature in these areas is the true master of creation.

The distinction between food and medicinal plants is fluid. Once taken for granted in plants, science occasionally records new medicinal effects; most fruits once considered very simply contain some of the world's most powerful anti-cancer agents. The same goes for tea-green tea is one of the most effective known anti-oxidants-and cinnamon spices that prevent the onset of weakening diabetes-and this list continues to grow. The same difference exists between natural fragrances and essential therapeutic oils. The rose oil, filled with citronellol, works three times as a perfume, as an active agent for herpes simplex virus, and as an uplifting scent that can help you open up emotionally after an endured trauma.

The use of so-called' alternative' therapy is growing, and more people turn to nature's wisdom to cope with all kinds of physical as well as psychological disorders. Ironically enough, from a naturopathic point of view, the root of most ills is out of balance with nature. Eat unnatural things and live in synthetic environments after irregular cycles. Lack of harmony with the world we are made of and live in leads to dissatisfaction in our minds and bodies. As the plants have created their marvelous botanical material in the process of "planting" our path back into balance by the rhythms of heaven and earth.

Depressive mental disorder is an essential reason for today's alternative and complementary therapies. A variety of physical, psychological, and energetic problems can lead to feelings of depression and an all-round view of negativity. However, trends can emerge once in a state of depression, making relief even more challenging to find. The use of essential oils to enhance psyche and spirit is becoming more common due to the extensive and dramatic effects of the oils. Although the oils themselves do not influence the cause of depression directly, they may help people release themselves from their depressed periods— they can provide an opportunity to' stop the sofa' so to speak, and make long-lasting changes. Most natural healers assume that depression is the result of being out of control, like so many other ills; they can not' align' with the natural state of equilibrium that permeates the world. Aromatherapy can be a compelling means to restore this equilibrium, to instill in your body and mind the most concentrated, pure natural botanical products.

For humans, the olfactive (smelling) area is roughly 2.5 centimeters for length and is found in each of the two nasal cavities between and under the ears. Compared to vision, we find that olfaction is much more complex-an, almost infinite number of component compounds can be identified at deficient concentrations. In order to experience the visible spectrum, people use only three forms of photoreceptor; the sense of smell, to comparison, depends on several hundred distinct receptor groups.

Psychological research has shown that natural plant oils activate many brain regions, including endocrine, limbic, and immune (emotional center) functions. Essential oils have a strong and profound effect on the complexities of the skin, emotions, and psyche. Essential oils have a clear and immediate impact on inhalation. When the capillary beds of the sinuses move, and olfactory nerves activate, resilient plants invade the brain and create clear and powerful systemic effects, which affect the emotions the most instantly. Our emotions and our sense of smell have strong ties-perhaps more than any other of the four senses.

Essential oils are known for enhancing prana flow (essential strength of life) in both natural and ayurvedic medicine, strengthening and nourishing the Ojas (energetic power and immunological essence), and enhancing Tejas (clearness and cognitive luminosity). Essential oils are remedies for Shen, the divine essence that lies in the heart and directs, and governs consciousness in traditional Chinese medicine. Consciously used, essential oils enhance positive mental and emotional conditions.

However, the medicinal effects of essential oils will help the heart and mind through their capacity to support physiological healing. In a Korean study on the effect of aromatherapy on pain in arthritis patients, pain AND mental depression were reduced significantly when massaged with marjoram, lavender, rosemary, eucalyptus, and peppermint oils.

Several oils that are known to lift improve prana, Shen, lighten Tejas, and cultivate Ojas, which may have significant effects on the symptoms of depression. Such oils can be used either alone or mixed in a nebulizing diffuser (which produces a fine nebula of inhalation oils), or in an aromatherapy massage, inhaled and absorbed throughout the skin simultaneously.

Bergamot (pressed from the skins of bitter oranges) is well known for its ability to raise gently. With regard to Chinese medicine, this results directly from the smoothing of the flow of Liver-Qi(' Chi' or Life Force') that the liver is seen as the seat of the eternal soul. Bergamot offers both the ability to relax and strengthen the nerves; it is ideal for various types of depression.

Neroli (bitter orange flor) controls Qi like Bergamot–and like jasmine floral oil, consoles mind and heart. The form of depression that comes from nervous and emotional fatigue is required at the core level of Neroli. Neroli elevates spirit and mind with its ability to feed and unify. Neroli helps to recover and release repressed emotions that can nurture was specifically designed for people that are isolated from their senses and emotions in order to escape psychological pain and suffering.

The Chamomiles (German and Roman) are ideal oils to use when the anxiety expresses itself in a moody, irritable, dissatisfied outward manifestation of stagnant liver Qi. These floral oils are earthy, rich, and grounded with subtle qualities of elevation.

If depression is (overly aggressive) of a Fire, often it includes a disequilibrium of joy and love–the heart and mind's core emotions. Joy is an extension of Shen's innate sense of harmony and perfection (spiritual essence), an experience of emotional and spiritual well-being. The heart and Shen depression include the lack of a normal sense of joy. There is often an underlying lack of enthusiasm and desire and a lack of creativity. Rose otto, or Rose Absolute, distilled rose oil, can have a profound effect on this condition. Rose is thought to open the first heart aromatically, bringing joy, raising, and restoring harmony.

The following are a few recipes to elevate and release depressed emotional conditions-use your intuition to find the right one. Often the most attractive essential oil or mixture you find is the one that best serves you. Experience, explore, and enjoy these fantastic natural gifts. Such blends may be used in a diffuser or candlelight unless a carrier oil is indicated-mixtures with carriers are specifically intended for massage through aromatherapy (self-massage is very effective, and a quick massage of a friend or loved one).

The Strategies on How to Declutter Your Life

Nobody truly needs to live their life loaded with clutter; however, for some, people clutter is all near. When clutter starts to develop around you, how to declutter your life ought to turn into a need. If you are not kidding about clutter-free living, you will begin to make legitimate moves to declutter your life. The steps essential to living clutter-free are not too entangled. Coming up next are useful systems that can show you how to declutter your life.

1. Before you start to declutter your life, it is critical to understand that touching base at your definitive objective of "no more clutter" ought to apply to all parts of your life. This not just implies you ought to declutter the space you live in; it likewise implies you ought to declutter your mind also. This is a significant idea to get a handle on because expelling clutter from all parts of your life evacuates obstacles that hinder shielding you from being profitable consistently.

2. To be effective in living clutter-free, you should start your decluttering process in the right mood. When you are feeling down or discouraged, it is easy to renounce the steps essential to keep clutter at the very least. It isn't unusual to find that clutter and sadness go connected at the hip. If this is your case, you should address the previous to invert the last mentioned.

3. When starting your decluttering process, I urge that you take things moderate and comfortable from the start. One of the most widely recognized errors people make is that they will attempt to take on an excess of too early. This leads to being overpowered and incapable of completing the undertaking. A fruitful strategy to landing at life without any clutter is to break the whole process into a progression of little errands and work on them individually. By breaking your general assignment into littler undertakings, you are probably going to accomplish an effective result.

4. Decluttering your life likewise means settling on hard decisions. Even though it might sound easy, settling on choices on what you are going to keep and what you are going to discard is a choice that shields numerous people from getting to be sans clutter. It isn't unusual for people to move

toward becoming pack rodents and clutch things they neither need nor need. For instance, a devoted peruser may wind up sparing a few books he/she doesn't care for! It just turns out to be natural to clutch these things without really thinking. When the opportunity arrives to state "No More Clutter!" you should free yourself of those things that have outlived their value.

5. Making a rundown of the considerable number of assignments you wish to handle when decluttering your home, office, and life is another significant advance. Indeed, an underlying agenda of the considerable number of things you have to sort out will help you in arriving at your objective. A confusing way to deal with the association is pointless. Pre-plan your methodology as most ideal as this will help your prosperity potential massively.

6. It likewise never damages to request help. Regardless of whether you approach one of your companions for a little guidance or you primarily read a "how-to" direct, you ought not to wade through the process without anyone else if positive outcomes are not prospective. Look towards accommodating counsel that can deliver incredible tips on taking advantage of your enemy of clutter system.

The process to make your life clutter-free isn't as confounded as you may be persuaded. You can understand your disorder by merely finding a way to turn around the issue. It is easy to do once you have the correct understanding of the process.

Are you one of those people that live a life loaded up with clutter. Have you at any point needed to declutter your life yet didn't have an inkling where to begin? To declutter your life, you should initially start by decluttering your mind. What number of you have perused a magazine chapter or a news story that appeared to drift from subject to theme with no coherency? Some may point to such composition as a successful type of Avante grade material; nonetheless, a great many people will see that composition for what it is, disorderly. Frequently, it is the consequence of a cluttered mind that needs a center and needs decluttering.

Presently, while you may notice such issues in others, would you be able to see issues such inside yourself? Typical indications of a cluttered mind incorporate absence of focus, neglect, outrage the board issues, tension, and eagerness. Fundamentally, the mind isn't working in the correct way it should. This leads to the previously mentioned conduct issues. That is why it is so useful to find a way to lessen mystic commotion and grasp the advantages of a decluttered mind.

This brings up issues concerning how to declutter your mind. The steps are not as intense as you may expect.

•Believe it or not, an absence of rest can prompt a severely cluttered mind. Some may figure they can perform successfully with little rest; in any case, reads have demonstrated that for a great many people, this isn't the situation. By going excessively long without the best possible measure of rest, you undermine your ability to have reasonable considerations.

•It appears as though you can't pass an accommodation store walkway without purchasing various caffeinated drinks. If you need to receive the rewards of a decluttered mind, avoid these kinds of stimulants. Abundance stimulant ingestion can prompt uneasiness and hustling contemplations, which can make a mind altogether cluttered. A calm mind is commonly not cluttered. Why ingest stimulants that undermine your ability to keep up a serene mind.

•Exercise is a typical method to declutter your mind. Some may find this somewhat difficult to accept and expect practice impacts the physical body. This isn't the situation as a sound body regularly sires a solid mind. Performing activities can diminish any uneasiness you might feel that causes issues and an absence of transparency in your reasoning.

•Engaging in a bit of loosening up stimulation can likewise calm the mind. It can even enable the brain to work all the more successfully. No, you would prefer not to squander hours sitting in front of the TV, tuning in to music, or perusing magazine chapter s, however putting some time into lackadaisical interests will counteract burnout and help in decreasing clutter of the mind.

•Increasing your necessary examination aptitudes can likewise declutter the mind. This should be possible in various ways. The process can go from something as elusive as examining progressed mathematic to just performing crossword baffles. Honestly, anything that hones the mind will make it more grounded and less cluttered. Among the numerous advantages of a decluttered mind is by and large rationally sharp and consistently on your game. That is why anything that fortifies the mind is well worth investigating.

•Calming and loosening up activities, for example, reflection and yoga are well known because of their effect on the mind. Such activities have been created through hundreds of years of development. That implies they have since a long time ago refined the ability to calm one's mind and wipe out potential clutter. Along these lines, why not incorporate such activities into your day by day routine.

•Eventually, the best of advantages of a decluttered mind is that you can live your life to the fullest without the mental issues a brain that needs calmness is known to epitomize. Instead of enduring in such a way, it is ideal to find a way to calm the mind and appreciate the advantages such a viewpoint would deliver.

Decluttering & Creativity

Instinctively the connection among decluttering and imagination bodes well isn't that right? Inventiveness flourishes in the place that is known for new ideas and open reasoning, while clutter will, in general, be portrayed by sticking on to old plans, dispositions, propensities, and assets. To free yourself up to be astoundingly creative, you frequently should be set up to relinquish the clutter first. Motivation is probably not going to develop except if you've made a space for it.

Clutter, for the most part, develops unobtrusively and intangibly after some time. The purpose behind this is that not all clutter starts its reality as clutter. If you consider the clutter in your life right now, you can likely perceive that quite a bit of it was initially valuable and significant. It's the progression of time and then moving on to different periods of your life that convert a considerable lot of your once-brilliant ideas, items, and connections into life clutter.

You'll most likely find that, for some odd reason, a portion of your old clutter comprises of items and ideas that were before your creative play area. A significant number of yesterday's creative sparkles develop into the present clutter. It doesn't imply that they weren't original at the time or that they had no value. That time has passed, and they are never again current. I like to envision them as the creative venturing stones that have carried me to where I am currently - I couldn't have here without them, yet their worth is presently before, and by sticking on to them, I keep myself from moving advances.

That is why decluttering must be a lifestyle, a perspective, and a continuous action, especially during the times when you need to deliver creative yield.

There's an essential qualification to be made among clutter and creative messiness, however. A peruser of my pamphlet kept in touch with me about her decluttering schedule: "I am a craftsman and in every case, clean my whole studio before starting another arrangement of artistic creations. Sometimes this may take two days! I put everything in the right place, vacuum, wash windows, revise the feng shui, and so on. When I am done, I favor the space and after that, continue to wreck it with all my creative materials and energy thoroughly!!!"

The space you declutter might be a physical or a psychological one - interestingly, it's reasonable, and that is the thing that enables it to be a creative start point. It frees you to get out the entirety of your shaded pencils, all your splendid ideas, all your intriguing words... to toss them in, blend them around and to make magnificently creative wreckage. At that point comes that stunning stream involvement of being wholly absorbed as, from the debris of creative potential, a feeling of concentrate slowly rises.

Most times, for me, I don't figure the center would come except if I permitted myself the creative messiness first.

In this unique situation, at that point, clutter is the stuff that squares you from having the unmistakable space in which to get creatively chaotic.

It might be natural clutter - physical things, social event residue, and occupying your creative room. That is commonly the most manifest sort of clutter to spot and to take care of.

Be that as it may, it might be mental or enthusiastic clutter. For instance: the inner voice that says you ought to continue ahead with something increasingly significant, or the dread of delivering creative yield that isn't perfect the first time. These idea examples and feelings are clutter as well.

To hold onto decluttering as a lifestyle and go it furthering your potential creative benefit, there are three essential abilities to create:

•Perceiving clutter before it even enters your life and halting it at source

•Recognizing which of your already helpful considerations, mentalities and items have now transformed into clutter

•Being set up to thank the clutter for its prior convenience, at that point let it go.

What life clutter might you want to thank for its handiness, before releasing it and liberating yourself up for new creative ideas and yield?

CHAPTER SEVEN

Establishing Order in Your Home & Life

Perhaps the snappiest approach to realize change and request into your life is to begin clearing your home of messiness. At the point when our homes are out of request, it appears in our timetables, our work, connections, and funds influencing all aspects of our lives. In my work, as an inside creator and holistic mentor, I've discovered the individuals who continually feel overpowered quite often have an issue with the mess in their lives.

Living with mess makes pointless pressure, disappointment, mayhem, and perplexity in your life. At the point when your house is all together with everything in its legitimate spot; your day by day schedules will stream easily. Envision when you need something how great it will feel to know precisely where it is. No considering it, looking for it, burrowing through heaps. It is actually where it is assumed to be. At the point when your house is free of messiness, and altogether it brings a true serenity and prosperity to your soul. Keeping your home mess-free and all together requires building up new propensities, applying new hierarchical techniques, and making new family schedules. It will need time, vitality, and inspiration on your part. When you have pledged to begin, you'll feel the impacts very quickly, and it's simpler to keep up the force.

The mess is the physical indication of inner clashes, which result in poor propensities, sloppiness, enthusiastic connections, and an excess of stuff. At the center, everything of messiness is a choice postponed. Request in your home moves toward becoming a request in your life. Those heaps of messiness that have been amassing in your carport, upper room, cellar, and extra rooms are influencing your growth. The heaps of messiness you have in explicit zones of your home have a relating impact in that aspect of your life.

At the point when things are strange, and you need to look through stuff to discover what you are searching for. It can and causes pressure and disappointment in your life; depleting your vitality. An assignment, for example, preparing the children for school, can transform into a strained encounter. Not having the option to see a bill that requirements as paid until the late charges have just been applied or the administration has been stopped. These are only a few instances of how mess can influence your everyday life.

The primary explanation we clutch things and amass massive heaps of messiness is a direct result of passionate connections. We've every shaped connection to the things we claim. Regardless of whether it's a negative or positive connection, we can get out and out regional about our thoughts. If you let go of your relationships, you would likewise relinquish the source, of quite a bit of your misery.

The relationship to our wants is about more than our material wants, yet additionally, our enthusiastic and profound desires. It's a relationship to how we figure things ought to be, or how we would honestly like them to be. There is a necessary arrangement. Keep your home condition mess-free, spotless, sorted out, and straightforward. Try not to pack your space, cupboards, or drawers with stuff you don't have utilization for. Keep just what you need and enrich only with things you adore. That's it!

Imagine your home and make a speedy rundown of the jumbled territories. To abstain from inclination overpowered, separate the review into little ventures that can be handed over some time, as opposed to one major undertaking to be done in a day.

Utilizing the three-box strategy is the best method to start clearing the mess. This strategy drives you to settle on a choice, thing by thing. You will require three boxes named "Store Away," "Give Away," and "Discard." Take the three boxes to the jumbled zone as you get each bit of messiness. Ask yourself, "Do I need it? Do I use it? Do I adore it" at that point, choose would I like to keep it, give it away, store it away, or discard it. Settle on brisk choices and spot the thing in the appropriate box. Once the crates are full; the store-away box is for those things that you can't settle on a choice about as of now, at that point, store them for a half year, if after that time you haven't missed the things in that case. At that point, settle on the choice to either give or discard the items. The give-away box ought to be expelled from your home inside five days. That leaves sufficient opportunity to make plans to drop the crate off at your neighborhood philanthropy or generosity. The cast-off box ought to be set out for the garbage pickup right away.

The less you have, the less you need to get, perfect, and set away.

The things that are left ought to be the things that are most critical to you and the things that you cherish. To have a deliberate house, you'll have to ensure there is a spot for everything in your home. Make sure to return things that you have utilized back to its particular area when you are finished using it. To keep the mess at the very least, practice this new guideline, one-in, and one-out. When you purchase something, some jeans, a shirt or book, you should dispose of something. For instance, if you are buying two shirts, when you return home, you should arrange two shirts. if you have a thing you have not worn in a year (a full season) odds are you won't wear it in the coming season. So choose to either give it away or discard it.

When you let go of things you've been clutching for quite a while, it discharges the enthusiastic connection you had to those things. By clearing the physical mess from your home, you additionally make the passionate, mental, and otherworldly confusion from your life. At the point when your house is all together, a fantastic remainder will pursue it. There will be a feeling of quiet and serenity in your life. You are free from dawdling, your life is streamlined, and you feel lighter, gain vitality, more prominent concentration, and inspiration to accomplish your life reason.

Why Frown In A Critical Situation

All want to know how to be happy. The problem is that few of us know how to be satisfied. Think for a minute about it. What do we think about when we believe we are so glad? For most of us, we think of living carefree, having more time and energy, being financially secure or prosperous, having the means to go somewhere or do something we've always wanted to do, finding our soul mate, or getting something we've always desired. When we think of these things, it feels good and let our minds wander a little as if we are "living" the visions in our heads. In a way, during these cognitive visualization periods, we are satisfied.

What happens to our daydreams and ideas about what would make us happy? Usually, unfavorable is the most immediate response. The first thing most people do is gage their new ideas against their current circumstances when they think about how to be happy. This is inadequate almost always because we always want to know, develop, and experience beyond what we already have. And we know how to be happy, but we feel helpless, furious, upset, deceived, and discouraged as if we were victims of a cruel world that doesn't care about us. There could be nothing beyond the facts!

You need to consider your thought processes and underlying thoughts about happiness if you want to know how to be happy. The reason is to understand and use your thinking's critical ability that has enabled everything you've ever received! Once you've seen the "evidence" for yourself, you'll be able to understand the fact that, regardless of your current circumstances, you can not only have anything you want but that you can always have!

Think of your house of dreams. Stand with the person you care most about side by side in front of it and respect it in your heart. Imagine what you'd feel like right now, just after you've bought this house. These are a few good feelings now! These are the emotions that make you smile regardless of where you are or what you do. Because these are the feelings, you are happy with. Now understand this, these are the feelings that will "attract" you to this house!

To learn how to be happy is to learn how to think productively. Get used to thinking about what you want as if you already have it, as in the above example. The focus should be on generating the most positive possible emotions of happiness and gratitude. So, as long as you can, HOLD those emotions. For many people, THINK of what they want will make them happy because they are conditioned by the current reality's negative emotional whiplash. They've spent so much time feeling depressed, disappointed, and angry that they don't have the means to buy what they want or the time to enjoy what they have, that they no longer spend the emotional energy in believing they're happy. They're lost, and life is cold, cruel, and hollow for them. I was there, and you've got chances as well. It is an unnecessary, sad, and hopeless state of mind.

The thing that most of us do not realize is that the universe reacts to all of our feelings! It's because we thought about a doughnut and had a good impression if we want a doughnut before we work. We searched our pockets mentally, of course, and had enough to spend on a doughnut. And our money was worth the good feelings of eating the doughnut, so we agreed to get it! Now all the way to work, we can feel good just thinking about that doughnut because we're "knowing" that we're going to get it very soon. We're just so happy that we can almost taste it right now!

That's how we draw it all. The drawback is that we immediately feel bad when we test our resources and don't have enough RIGHT NOW to get it. You are sending out a message to the world saying just that by feeling angry or sad or thinking that you "don't have enough" or "don't deserve it" or "don't get it." You use the same power to attract a basic doughnut to repel what you most want! Through this mistake, you attract the circumstances through the law of attraction that gives you precisely what you tell the universe you wish to, which is always and most often what you think best for.

When learning how to be content, note that what you are emotionally sending out is precisely what you are going to get. First, you've got to give. Give feelings of appreciation and joy. Find something you can feel good about and concentrate on strengthening those emotions and enjoying them. All the while, assured that fate wheels work hard to bring you MORE of stuff that makes you feel good!

It's not a challenge to know how to be happy but an option. You can choose to have the confidence to feel and own in your heart what you want most now. Or be scared of what you most desire because of how terrible the emotional reaction when you do it will be. They are nothing more than irrational "temper tantrums," and you will avoid them with greater and greater ease as you grow mentally and emotionally. Just by realizing that these negative emotions you've always experienced in reaction to "not getting enough" thoughts are what has held you back, the way you use your mind for the right will shift.

It takes practice to shift the mental focus, but it is gratifying. Dare to dream and imagine the fulfillment of all your wishes and aspirations from now on. Reflect on your primary emotions, those that bring forth the most potent and enduring feelings of joy, satisfaction, and anticipation. I refuse to think about them negatively. Always be mindful of the main point of failure you've been slipping into in the past so many times. The bottomless pit of negative emotion where you have lost so many of your wildest dreams. Build a bridge over the gulf and focus on feeling good about what you want! Even if you don't have it already! Also, if you don't also have a HOW CLUE, you're going to get it! Know that you're going to believe it and that you can taste it already! Just like that doughnut, without a doubt, it was yours emotionally even before you put it in your pocket. At first, it may sound crazy;

it may even be painful. Be mindful of your feelings, choose to feel good, feel grateful, and want to replace negative emotions with even more positive emotions.

Don't want something to make you happy. Instead, choose to focus on being content NOW. Try and use these "happy thoughts" to feel good all the time. They're going to lift you to the heights you can imagine. Understand that your present circumstances are simply the result of your past thinking and feeling. You can choose every second of every day to change that. So you won't have to ask how to be happy anymore. Because in your life you should draw more of what you want! Right now, by feeling good about it, you can start attracting what you want most! No matter what, those emotions that make you smile are the arms that you need to master. They are the lights you'll use to push away from your mind the limiting shadows, revealing the limitless power of every strong wish. The bottomless pit of negative emotion where you have lost so many of your wildest dreams. Build a bridge over the gulf and focus on feeling good about what you want! Even if you don't have it already! Also, if you don't also have a HOW CLUE, you're going to get it! Know that you're going to believe it and that you can taste it already! Just like that doughnut, without a doubt, it was yours emotionally even before you put it in your pocket. At first, it may sound crazy; it may even be painful. Be mindful of your feelings, choose to feel good, feel grateful, and want to replace negative emotions with even more positive emotions.

Don't want something to make you happy. Instead, choose to focus on being content NOW. Try and use these "happy thoughts" to feel good all the time. They're going to lift you to the heights you can imagine. Understand that your present circumstances are simply the result of your past thinking and feeling. You can choose every second of every day to change that. So you won't have to ask how to be happy anymore. Because in your life you should draw more of what you want! Right now, by feeling good about it, you can start attracting what you want most! No matter what, those emotions that make you smile are the arms that you need to master. They are the lights you'll use to push away from your mind the limiting shadows, revealing the limitless power of every strong wish.

What Prevents Genuine Happiness

The world lacks peace, and we waste an excessive amount of time striving to be happy. We do this in many forms, such as music, party, etc. Such stuff makes us happy on the ground momentarily.

The joy we are looking for is deep inner happiness. External happiness is OK, but we are not happy without inner happiness. Perhaps what we're looking for is happiness. We're looking for joy, but we say we're looking for happiness because the smile on our face can calculate happiness. In our fast-paced artificial world, comfort is more in tune. Joy does not seem to suit somehow. So how do we get pleasure and satisfaction? Happiness and joy are not things that you get. You are a natural result of action in a certain way, that is to behave according to specific rules, and you will be satisfied with the course, it is a guaranteed result.

So why aren't you pleased? The response is you've heard specific rules or instructions. Indeed, there are laws. You've learned the rules of how to behave. We have, of course, brought about the way you are now. To be happy, you need to change your learning. You will change the way you feel if you change your knowledge. Improving your learning to those that will naturally lead to what you want, happiness and joy, is therefore significant. Adjust your reading, and inevitably the outcome will occur. You are becoming a happier and fulfilled man. It sounds so simple, and that's it. The opposite is exact as well, it's difficult and complicated, that's the way with human nature, it's a combination of opposites, and the art of living well is to find the balance of those opposites.

Or put it in the simplest terms:-Behave with your inner self (soul) in conformity and peace, and you will be satisfied. What makes it impossible for us to act and behave and feel like our soul? Adherence to other things. That's it!

Small reflections of your heart are the most significant moments of joy. Time stood still when you just experienced the moment and forgot yourself until you saw a glimmer of what shines deep inside you. You've been disconnected from everything at that moment, and you've been lost in the moment. You've actually lost your identity for that moment. You haven't been there! You, your definition wasn't there, and the feeling was one of reality, a reality that is always there but out of the consciousness of self.

The more harmonious we are with our soul, the more dignity we have, the more relaxed and full we feel, the more happiness we feel. Satisfaction is a soul-harmonizing human feeling. Sometimes when we listen to music, we can feel joy when the music feels real when we hear the music and when we become one with the music. We are disconnected from ourselves and everything else, music is there, we get lost in it, and we forget ourselves.

Stress and stress disappear in these moments, and our physical body is resting and healing, thankful for this precious time. So we can quickly detach, it's hard for us to separate. The problem is that when we can disconnect, we have so few conditions or choices.

Attachment for something else! Unhappy. It's the formula. Does the soul's devotion bring happiness? No! Let's uncover a little bit of the puzzle, and the soul is what it is. It's just that. It's a beautiful thing, like beautiful scenery, and like beautiful scenery, we can bathe in it, feel in harmony with it, feel the essence of beauty. We feel joy when we do, and we think the attraction we are. We believe that our creation is magic, beauty. If we get addicted to the scenery, we want to own it, own it, and command it. Most of us aren't addicted to the view, it seems like a stupid idea, and we aren't, because the scenery and nature are too high. We own landscape paintings.

Trying to own, possess, and influence the soul merely is denying what it is, so attachment to the soul is not going to bring happiness. To be in harmony and align with the will of the heart, like the will to enjoy the scenery.

Then the formula is Attachment= Unhappiness. Indeed, two questions arise, why are we connecting, and why is it equivalent to (creating) unhappiness? Why are we adding it? If I tell that myself, good question. Look at the ego. The identity of the ego causes duality, and personality comes from nothingness, only a thought, not structure, but an idea. Yeah, but what an experience and what a real idea it sounds. If it sees the truth and beauty of the soul when identity is first created (in childhood), then it may want to be one with soul, and if it does (in its infant way of thinking), it fears to return to nothingness and to cease to exist. Sheer terror consumes the fragile ego. And thus, the solidification of fear begins, fear of one's weakness, so starting without foundation. Look in the other direction. Quick, look out, look at your mom, she's holding you in her arms, and you're safe, and now you're, she's calling you by a name, and I'm that.

So why are we adding it? I'm asking you what we own, why we want more, why we own? We want to own things, objects, material things that we see, real tangible objects that can be marked, and we want to feel them, touch them, show them off, let others know we have these things, we want to be like them, reliable, with real shape appearing. The better they have, the more. The better the situation we have, the more genuine I am. These things are mine, always trying to prove that I am, I am real! They're part of me, I own them, I control them, so I'm healthy, and if I'm strong, I'm not vulnerable. And I don't have to blame myself if I'm not weak. And if I can manage people, oh, how strong that is! I am!

We connect because we are vulnerable, but unfortunately, attachments only seem to make us feel heavy. We feel safe, powerful momentarily, and only when others are conscious of it. We've got to keep doing what we're doing, so notice others. We're trying to prove ourselves to others because we don't believe it on our own, we can fool ourselves, but we don't think it deep down. The more we have and the more (so-called) dominant, we become, the weaker we become because the power depends on that power being socially recognized. It is, therefore, easily lost, and the risk increases as such. Through connecting, the personality seeks to thrive; it's an endless battle because it's a negative feedback loop.

The method is the following:-

1. Vulnerability status. It feels insecure— anti-survival.

2. It needs to survive.

3. How? Be against the weak-powerful and heavy.

4. Attach-Own, hold power.

5. Ego/identity feels reliable and robust (vulnerability relieves tension).

6. The feeling easily subsides and returns from vulnerability.

7. Vulnerability status and so on, and so on.

So, why does attachment mean unhappiness? Happiness is a good state of everything, a free state of stress, a state in harmony with the soul. This state is an actual state. It feels complete and whole, a pleasure to be. The country does not need propping up or help, and it does not need something to be done; it has no needs, it is what it is. So what happens if we attach? If we get attached to an object, that means the objective is essential, because it is worth connecting to. The conscious brain processes information linearly and like a computer, it processes as a sequence of yes or no, and on or off, computers process as either 0 or

1. The brain and computer process incredibly fast, which makes it seem as we can process a lot of information simultaneously. But it is sequential on or offs, this or that's.

We have an entity that we are connected to. The object is important. We are attached to the object, so it must be relevant. We, the ego, naturally feel (attachment) the purpose is valuable and hence necessary. What is not eligible is how vital, more important than what? Less relevant than what? There are no such credentials. The ego accepts the critical thing is the artifact. The ego creates this assertion so that it can bind to the object to withstand weakness for its intent.

The processing that takes place during attachment 1 is below. Want to own it (no 4 above).

2. It is necessary to have an artifact.

3. Get the artifact.

4. There's an implicit order.

5. Unconscious accepts or dismisses command.

6. Order of unconscious processes (if accepted).

7. The state of the ego is polluted.

8. Debriefing. Unconscious processes the following information.

9. It is necessary to have an artifact.

10. There are the ego and the object involved in the event (the order from ego is something like" "I want that").

11. The goal is critical-the ego is insignificant. (Duality, value is subjective. Therefore the ego puts a priority on the object through choice, the processing brain only has two representations of the object and the ego, it recognizes the order (in this case there are no pre-existing conditions to deny it) the purpose is immediately marked as essential, or the law would not be processed. There are no shades of gray to the sensory brain characteristics. There's something significant or not; it's either necessary or unimportant. Outstanding quality is either on or off. Even if something else is irrelevant, can something be important? The ego and the object are attached, and the two are connected, the one has shifted importance, the other has shifted emphasis! This is the price of terms and tags; this is the only manner in which this data can be stored, as soon as the ego uses the critical and useful tag. The ego will likely be branded as unimportant or useless (unconsciously).

12. The debriefing order and resolution are complete. There is no feeling or sense that the data is being processed.

13. The knowledge that the ego is insignificant is expressed as a sensation and returned to the ego and our heart. It's translated into a feeling for feeling is an experience, and our inner world's language and words are, well, they're just words, meaningless vacuums until you put in whatever meaning you'd like to put.

14. We have somewhat sullied our soul, and the ego has more evidence to support its negative beliefs.

When we know that attachments cause unhappiness, we have a choice, and do we want to be happy or unhappy with the decision? It's the ego that has the power of choice as I say we and the ego has a problem. It feels vulnerable and has a strong desire to minimize this insecurity, which is understandable, of course.

We will make smarter decisions with experience and accurate information if we choose to be healthy. Then we need to know what makes us happy and unhappy. We now know what makes us sad, and remaining attached to attachments is futile.

As we bind to things, we spend a part of us, so each time we do, it is fragmenting and weakening us instead of strengthening us. You may be afraid that you need to get rid of everything to be happy. No, you have to remove yourself. To be disconnected does not mean to be in a state of non-care. Feeling disconnected means unimportant are the things you have. They may be helpful and beneficial, but if we are detached, it means that we are not emotionally attached to them, we are not invested in them. We are what they are, material objects, and we, human beings, are what we are. We are useful items and are not essential artifacts of content. When material objects are more full, more important than our egos, we will feel less insecure.

We connect our identity to objects, people, thoughts, values, and the most troublesome attachment. We have invested so much energy in our character and feel that detachment from our identity is impossible. We believe that we are our selves. Nevertheless, status is a theory, an idea, a precious idea, but no less an idea. In fact, by attaching ourselves to our identity, we deny our reality. The truth is that we are a physical form, and we live on Earth on this earth, and we have a blueprint of how to survive. This model represents the universal essence, and it is our essence.

Every person on the planet has the same blueprint, our unique fingerprint, how we express that blueprint. We don't see the design in others when we invest in our identity. In others, we don't see humanity, because identity is' I'm not you.' If we fail to see morality in others, we only see their actions, we reject their nature, their design, and we deny our own because they are the same.

Our identification is useful and helpful, but our weakness is also induced by it. When we remove ourselves from reality, that is, see it for what it is, a part of us. Then we can be following our hearts. We're going to know real happiness. When we let go of things, thoughts, memories, if we choose but don't need them, we can have them. We realize our power, we have confidence in ourselves, and we do not need control over others.

We're going to respect everybody. We're going to love all humanity. In everybody, we will recognize the human spirit. While we see actions are very often not following that spirit, and we can see that action is not the individual and that ignorance is reflected in bad behavior. The more we choose to be like our essence, the more we will be fulfilled, and happier our lives will be.

CHAPTER EIGHT

Decluttering Your Home

Home is something for all of us like a comfortable nest. It has to be clean, spacious, and well organized to provide us with relaxation and to nurture a favorable perspective. But when you have lots of unwanted items all around the house, your healthy life will undoubtedly be hampered. It's time to think about your home, decluttering.

Many people find the very concept of living a streamlined life with fewer tools appealing. Yes, we understand the advantages of getting fewer belongings, which implies less organization, less washing and cleaning, less debt, and fewer headaches. The extra time and money saved is spent on some of the biggest passions. If you're prepared to declutter, issues may emerge next time-where should I begin? The job is not painful if the trick is known.

There are a variety of creative and exciting ways to participate smartly.

1. Plan a list

Do not believe the whole building is; instead split it into sections and create a list of locations to be organized more. This does not mean only spaces such as a living room, children's bedroom, master bedroom, etc. but also

103

wardrobes, bookshelves, TV rooms, kitchen clothes and more.

2. The gift rule

Get prepared to depart from certain items first. Trust me, and you'll never want them back once they're gone. Give every day one post. Contain just one job but make sure it is a daily practice. Continue for three months. In the end, you will understand how clean and comfortable it feels at home.

3. A must-have trash bag

You must always have a trash bag, and it should be large enough to grab whenever you want. One good recommendation is to maintain a unique basket type item worth putting inside the house. There's no way to forget the task of decluttering home when you move through this basket many times a day.

4. Experimenting with a fun number

Choose a number such as 10, 12, 14, or 15. Choose 10 to throw away, ten items to give, and 10 items to move when you collect unwanted products. Whatever the amount may be, adhere to it and obey the same 10-10-10 or 14-14-14 rule. So many things are organized in a short time

5. The challenge of "living with less" is to be experimental

Yes, I'm speaking of the minimalistic lifestyle. Choose 30 clothing you're very attached to and wear for only three months those restricted clothes. The concept does not sound so pretty at first, but gradually you get used to it. After that, you can also attempt the same thing with kitchen appliances. It teaches people to live with less.

6. Follow the three-box method

Sometimes crazy thoughts are great! Take into account the three-box technique; you only need three large cartoon boxes before you begin to clean the chaotic rooms. One is for garbage, one for a donation and the latter for relocation. Start from a single room and go to the whole house.

How to Start a Minimalist Business

If you look at your local bookstore's company section or browse the Internet, you will discover a lot of data on how to start a company. Looking at some of them to-dodo lists, it's enough to disappoint even the committed person who wants to get started alone. Write a business plan. A business plan. Get a tax number. Get a tax number. Hire a reasonable attorney for the company. Decide on your company's legal framework. Research your market and your content thoroughly. Choose your niche in your niche. Get your training. Get your training. Hire an accountant. Hire an accountant. Register your company name. Register your company name. Get all your permits and licenses. Get your company funding. And the list continues and continues.

Sigh. It is no wonder so many individuals say "one day" they're going to begin a company, but never.

This doesn't mean that many items that are not essential in these lists. Some of them are crucial to your business ' success as you gain momentum. But most of these duties are not essential at the start. Even worse, they can become so overwhelming that they can paralyze you with indecision and distract you from the activities that make you money, the most significant tasks in your company.

Let's examine the five measures that a minimalist could take to start a company.

1. Decide for yourself what you love to do.

Passion is what will take you through the worst of your company. If you don't like what you do, when things get tough, you will give up and seek something more comfortable, which will get you the same money. If you're not enthusiastic about what you do, you're continually looking for something you love. You will never be as pleased as you can be so that your environment can readily be distracted by the bling. So why not begin your company on the correct foot around your passion?

2. Decide on a Product or Service

It is difficult to think of a passion that could not contribute to the growth of a product or service. Love math? Love math? Tutoring services can be provided. Do you love to give guidance? You can be a trainer (or a blogger). Love to hike? Create a local walking map or sell tours. Love to read? Love to read? Make yourself a critic. Love to muck up? Love to muck up? Become an artist on the road. Love meditating? Learn a meditation course or create an audio item for meditation. Your imagination only restricts the range of products and services that you can offer depending on your enthusiasm. Start with one or two and over time, build on it.

3. Figure Out the Money

There is no escape from your business mathematics. Even at its most fundamental level, you have to ensure that you charge your products and services more than you cost to generate them. Otherwise, it is only a hobby, not a company. Figure out the prospective money flow in advance can assist you in deciding which goods and services are most feasible for your minimalist business ' achievement. For instance, if you like Quilt, and the market only pays $400 per Quilt, you will see that you only make 5 cents an hour to create a Quilt, you may have a distinct concept based on your enthusiasm, for instance by learning quilting. Some ideas need only be slightly changed to make them profitable. For example, if you would like to teach meditation, you could attempt marketing directly with your colleagues by phone or email, instead of creating costly posters and running the course in your cellar. You'd be lucrative with the first student in this manner. And then you can construct it.

4. Figure out how you tell people about it.'

You need not be a genius to market your company. I wasted some years of money and gave so-called expert marketing tips, and all I did was pocket-line. Use only the communication lines you have already in place to get the word out at the start. If you use email, individuals that you understand will receive the email. Tell them what you're up to if you've got Facebook friends. If you use your phone for a while, call your friends or text and ask for assistance to get your message across. Be sensitive and beware of marketers who say they've got a magic bullet. Marketing is about relationships, so begin with those you have already.

5. Take action

Once you have determined what you like, what products and services you are going to offer, how you are effectively going to create cash and how you are going to get the word out, it's time to begin to take measures for your own company. Have a beautiful stretch and then work. You only have a sustainable company if you make money, so you feel like a minimalist company when you begin to see the cash flow. And proudly wear your badge.

Life Happiness Depends on Your Life Goal

Someone doesn't know what course is better for them. Someone doesn't know what shirt or clothes to choose from. Someone doesn't even know what to eat for lunch. Everyone likes to ask for a different opinion about their own question. Yet their life continues to repeat all this without worrying about solving them. Why do people find it hard to make a simple decision? People thought it was a life for an ordinary human being?

Actually, life is very easy. It all depends on your purpose in life. If you don't have a goal for life or don't have a clear idea about your goal for life, your life isn't simple anymore. You are not sure about your life goal if you don't know which path is right for you. If you are asking other people for advice in your own life (not relating to knowledge), you are not sure about your life purpose.

I firmly believe you should tell yourself what your life goal is when you don't know which shirt you should wear today. If you're upset with someone, tell yourself what's your life's goal. If you are concerned about the outcome of your test, you must tell yourself what your life goal is. We ought to be responsible for our own emotions. You should tell yourself what is wrong with you if you find you are in a kind of emotion that you don't like. Isn't your life goal all right? And, even worse, you have no clear objective.

Why can we allow another issue to occur? Why don't we try once in our life to solve it? We assumed that the issue was usual for an ordinary person? Did we think life was packed with many unfortunate events/things? We don't seem to be responsible for our feeling.

If we have set our life goal, we should be responsible for all its consequences, including our anger. If we can't accept these effects, we're expected to change the goal. The target could make our lives exciting. The target could always make us fail. The target will make our life simple and happy. What kind of life you all want depends on your purpose? Please do not blame people for being sad. It's really ridiculous to ask for advice on our own life. Why people can help if they don't know your life goal and you only know your own life goal. Someone even offers guidance on their own purpose in life rather than yours. They thought you should also use the same approach when you solve the problem.

Let's say you're going to choose a university education, and there's course A and course B ahead. If you choose course A, are you going to your goal? Please ask yourself. If you choose course B, can you follow your objective? If both courses allow you to achieve your goal, choose one. If you just take Class A to achieve your goal, choose Class A. If you can not achieve your goal in both courses, you must find the course that helps you to achieve your goal. What's so hard? The hard part is if we have no life goal. If that is the case, no one can rescue you. You can't ask someone how the problem can be solved. You are the nature of the problem. It's not from the setting. There are, of course, some people who solve the problem through another process. You try to find out which course will be more popular or which course will give you more money in the future. You no longer have this kind of concern (internal factor). It's an outside factor. If you have incorrect information, notice it and feel angry about it after you enter the course, what triggers your anger? It is caused by you, not by the person who gave you the data. When choosing your course using this type of process, you should be mindful of this type of risk and consider all consequences.

Happiness is personal. Happiness is personal. So we should do whatever we do, make whatever choices, and not let anyone do this for us. We must base our decision on what we want for our lives. We can't let others decide (just my suggestion) our happiness. When you allow others to decide for you, you have to accept the consequences and be content with everything you face. It's difficult to find someone who allows others to control their lives and still feel very happy with their lives.

The primary factor in differentiating adults with children is that adults can take responsibility and know the consequences very well, but children can't. When we look around, we can find many people with an adult body who cannot take responsibility for the consequences and do not know or care about it. You can't be responsible for their own emotions in general. In reality, our emotion is the product of our decision.

Our goal of life is our option. We should think deeply about achieving our best goal. We have 100% control to choose our own life goal based on our purpose and what we want. After the target has been determined, we work for our goal. Everything that happens after that affects our emotions. Emotion is, therefore, the product of our purpose in life. Emotion is our own option. Emotion.

CONCLUSION

How do we succeed in life and depend on the power of self-belief? Can you really just believe in yourself and let go of constant doubt, anxiety, and fear of failure? How can you learn to stop thinking to help you succeed? These are areas of concern for both the creative and adult and the family lives, and I hope this section will allow you to see the inner light that genuinely wants you to trust in your ideas and your approach to life. Let us do this by considering, for example, why a small business faces self-confidence that is key in making the right choices. You could be like other people and start businesses just to believe in yourself. Excessive thought can kill any chances of the impact of self-belief. There are important considerations. It is important to analyze every circumstance of your daily life before we approach it, but we too often swirl around a lake of psychological stress and confusion because we overthink. The overthrowing issue leads to future regrets and missed opportunities, but it can also be overcome to make the most of your life.

Are you somebody that likes to overthink stuff? Nonetheless, what exactly is overthinking? This book explains precisely what overthinking is and how it can stop your efficiency, productivity, and mood from being reduced. Do not allow overthrow to interfere with your ability to take the requisite risks, to help others, and to make the most productive efforts.

We must always be mindful of our thoughts and actions, but a person may influence himself negatively by overthinking issues in which they are obsessive or compulsive in a manner that leads to self-destructing or self-defeating behavior. Very often, the person does not really know how to solve these problems and can revert to envy and denial that nurtures its insecurities.

CPSIA information can be obtained
at www.ICGtesting.com
Printed in the USA
BVHW090513110521
606943BV00005B/1360

9 781802 165647